This Bee Keeping Journal & Log Book Belongs To

____

COLONY NAME	
DATE	
TIME	

WEATHER CONDITIONS					
___	☀	⛅	🌧	⛈	❄
___	☐	☐	☐	☐	☐

INSPECTION						
HIVE NUMBER	1	2	3	4	5	6

PRODUCTIVITY & REPRODUCTION						
AMOUNT OF HONEY						
GENERAL POPULATION						
AMOUNT OF BROOD						
AMOUNT OF SPACE						

BEHAVIOUR & ACTIVITIES						
USUAL ENTERING AND EXITING ACTIVITY?						
CALM BEHAVIOUR WHEN OPENING HIVE?						
BEES BRINGING POLLEN INTO HIVE?						
SIGNS OF ROBBERY AMONG THE BEES?						

HEALTH STATUS						
BEES SEEM WEAK OR LAZY?						
HIGH AMOUNT OF DEAD BEES?						
QUEEN BEE IS PRESENT / IDENTIFIABLE?						
INFESTATION BY ANTS / ANTS PRESENT?						
INFESTATION BY WAX MOTH / WAX MOTH PRESENT?						
NEGATIVE ODOUR NOTICEABLE?						

COLONY NAME	
DATE	
TIME	

WEATHER CONDITIONS						
Temperature	___	☐	☐	☐	☐	☐
Wind	___	☐	☐	☐	☐	☐

INSPECTION

HIVE NUMBER	1	2	3	4	5	6

PRODUCTIVITY & REPRODUCTION

	1	2	3	4	5	6
AMOUNT OF HONEY						
GENERAL POPULATION						
AMOUNT OF BROOD						
AMOUNT OF SPACE						

BEHAVIOUR & ACTIVITIES

	1	2	3	4	5	6
USUAL ENTERING AND EXITING ACTIVITY?						
CALM BEHAVIOUR WHEN OPENING HIVE?						
BEES BRINGING POLLEN INTO HIVE?						
SIGNS OF ROBBERY AMONG THE BEES?						

HEALTH STATUS

	1	2	3	4	5	6
BEES SEEM WEAK OR LAZY?						
HIGH AMOUNT OF DEAD BEES?						
QUEEN BEE IS PRESENT / IDENTIFIABLE?						
INFESTATION BY ANTS / ANTS PRESENT?						
INFESTATION BY WAX MOTH / WAX MOTH PRESENT?						
NEGATIVE ODOUR NOTICEABLE?						

COLONY NAME	
DATE	
TIME	

WEATHER CONDITIONS						
Temperature	____	☐	☐	☐	☐	☐
Wind	____					

INSPECTION						
HIVE NUMBER	1	2	3	4	5	6

PRODUCTIVITY & REPRODUCTION						
AMOUNT OF HONEY						
GENERAL POPULATION						
AMOUNT OF BROOD						
AMOUNT OF SPACE						

BEHAVIOUR & ACTIVITIES						
USUAL ENTERING AND EXITING ACTIVITY?						
CALM BEHAVIOUR WHEN OPENING HIVE?						
BEES BRINGING POLLEN INTO HIVE?						
SIGNS OF ROBBERY AMONG THE BEES?						

HEALTH STATUS						
BEES SEEM WEAK OR LAZY?						
HIGH AMOUNT OF DEAD BEES?						
QUEEN BEE IS PRESENT / IDENTIFIABLE?						
INFESTATION BY ANTS / ANTS PRESENT?						
INFESTATION BY WAX MOTH / WAX MOTH PRESENT?						
NEGATIVE ODOUR NOTICEABLE?						

COLONY NAME	
DATE	
TIME	

WEATHER CONDITIONS

	☐	☐	☐	☐	☐
___					
___					

INSPECTION

HIVE NUMBER	1	2	3	4	5	6

PRODUCTIVITY & REPRODUCTION

AMOUNT OF HONEY						
GENERAL POPULATION						
AMOUNT OF BROOD						
AMOUNT OF SPACE						

BEHAVIOUR & ACTIVITIES

USUAL ENTERING AND EXITING ACTIVITY?						
CALM BEHAVIOUR WHEN OPENING HIVE?						
BEES BRINGING POLLEN INTO HIVE?						
SIGNS OF ROBBERY AMONG THE BEES?						

HEALTH STATUS

BEES SEEM WEAK OR LAZY?						
HIGH AMOUNT OF DEAD BEES?						
QUEEN BEE IS PRESENT / IDENTIFIABLE?						
INFESTATION BY ANTS / ANTS PRESENT?						
INFESTATION BY WAX MOTH / WAX MOTH PRESENT?						
NEGATIVE ODOUR NOTICEABLE?						

COLONY NAME	
DATE	
TIME	

WEATHER CONDITIONS						
___						
___	☐	☐	☐	☐	☐	

INSPECTION						
HIVE NUMBER	1	2	3	4	5	6

PRODUCTIVITY & REPRODUCTION						
AMOUNT OF HONEY						
GENERAL POPULATION						
AMOUNT OF BROOD						
AMOUNT OF SPACE						

BEHAVIOUR & ACTIVITIES						
USUAL ENTERING AND EXITING ACTIVITY?						
CALM BEHAVIOUR WHEN OPENING HIVE?						
BEES BRINGING POLLEN INTO HIVE?						
SIGNS OF ROBBERY AMONG THE BEES?						

HEALTH STATUS						
BEES SEEM WEAK OR LAZY?						
HIGH AMOUNT OF DEAD BEES?						
QUEEN BEE IS PRESENT / IDENTIFIABLE?						
INFESTATION BY ANTS / ANTS PRESENT?						
INFESTATION BY WAX MOTH / WAX MOTH PRESENT?						
NEGATIVE ODOUR NOTICEABLE?						

COLONY NAME	
DATE	
TIME	

WEATHER CONDITIONS						
Temperature	___	Sunny	Partly cloudy	Rain	Storm	Snow
Wind	___	☐	☐	☐	☐	☐

INSPECTION						
HIVE NUMBER	1	2	3	4	5	6

PRODUCTIVITY & REPRODUCTION						
AMOUNT OF HONEY						
GENERAL POPULATION						
AMOUNT OF BROOD						
AMOUNT OF SPACE						

BEHAVIOUR & ACTIVITIES						
USUAL ENTERING AND EXITING ACTIVITY?						
CALM BEHAVIOUR WHEN OPENING HIVE?						
BEES BRINGING POLLEN INTO HIVE?						
SIGNS OF ROBBERY AMONG THE BEES?						

HEALTH STATUS						
BEES SEEM WEAK OR LAZY?						
HIGH AMOUNT OF DEAD BEES?						
QUEEN BEE IS PRESENT / IDENTIFIABLE?						
INFESTATION BY ANTS / ANTS PRESENT?						
INFESTATION BY WAX MOTH / WAX MOTH PRESENT?						
NEGATIVE ODOUR NOTICEABLE?						

COLONY NAME
DATE
TIME

WEATHER CONDITIONS						
	___					
	___	☐	☐	☐	☐	☐

INSPECTION

HIVE NUMBER	1	2	3	4	5	6

PRODUCTIVITY & REPRODUCTION

AMOUNT OF HONEY						
GENERAL POPULATION						
AMOUNT OF BROOD						
AMOUNT OF SPACE						

BEHAVIOUR & ACTIVITIES

USUAL ENTERING AND EXITING ACTIVITY?						
CALM BEHAVIOUR WHEN OPENING HIVE?						
BEES BRINGING POLLEN INTO HIVE?						
SIGNS OF ROBBERY AMONG THE BEES?						

HEALTH STATUS

BEES SEEM WEAK OR LAZY?						
HIGH AMOUNT OF DEAD BEES?						
QUEEN BEE IS PRESENT / IDENTIFIABLE?						
INFESTATION BY ANTS / ANTS PRESENT?						
INFESTATION BY WAX MOTH / WAX MOTH PRESENT?						
NEGATIVE ODOUR NOTICEABLE?						

COLONY NAME	
DATE	
TIME	

WEATHER CONDITIONS						
Temperature	___	☐	☐	☐	☐	☐
Wind	___	☐	☐	☐	☐	☐

INSPECTION						
HIVE NUMBER	1	2	3	4	5	6

PRODUCTIVITY & REPRODUCTION						
AMOUNT OF HONEY						
GENERAL POPULATION						
AMOUNT OF BROOD						
AMOUNT OF SPACE						

BEHAVIOUR & ACTIVITIES						
USUAL ENTERING AND EXITING ACTIVITY?						
CALM BEHAVIOUR WHEN OPENING HIVE?						
BEES BRINGING POLLEN INTO HIVE?						
SIGNS OF ROBBERY AMONG THE BEES?						

HEALTH STATUS						
BEES SEEM WEAK OR LAZY?						
HIGH AMOUNT OF DEAD BEES?						
QUEEN BEE IS PRESENT / IDENTIFIABLE?						
INFESTATION BY ANTS / ANTS PRESENT?						
INFESTATION BY WAX MOTH / WAX MOTH PRESENT?						
NEGATIVE ODOUR NOTICEABLE?						

COLONY NAME	
DATE	
TIME	

WEATHER CONDITIONS						
	___					
	___	☐	☐	☐	☐	☐

INSPECTION						
HIVE NUMBER	1	2	3	4	5	6

PRODUCTIVITY & REPRODUCTION						
AMOUNT OF HONEY						
GENERAL POPULATION						
AMOUNT OF BROOD						
AMOUNT OF SPACE						

BEHAVIOUR & ACTIVITIES						
USUAL ENTERING AND EXITING ACTIVITY?						
CALM BEHAVIOUR WHEN OPENING HIVE?						
BEES BRINGING POLLEN INTO HIVE?						
SIGNS OF ROBBERY AMONG THE BEES?						

HEALTH STATUS						
BEES SEEM WEAK OR LAZY?						
HIGH AMOUNT OF DEAD BEES?						
QUEEN BEE IS PRESENT / IDENTIFIABLE?						
INFESTATION BY ANTS / ANTS PRESENT?						
INFESTATION BY WAX MOTH / WAX MOTH PRESENT?						
NEGATIVE ODOUR NOTICEABLE?						

COLONY NAME	
DATE	
TIME	

WEATHER CONDITIONS						
Temperature	___					
Wind	___	☐	☐	☐	☐	☐

INSPECTION						
HIVE NUMBER	1	2	3	4	5	6

PRODUCTIVITY & REPRODUCTION						
AMOUNT OF HONEY						
GENERAL POPULATION						
AMOUNT OF BROOD						
AMOUNT OF SPACE						

BEHAVIOUR & ACTIVITIES						
USUAL ENTERING AND EXITING ACTIVITY?						
CALM BEHAVIOUR WHEN OPENING HIVE?						
BEES BRINGING POLLEN INTO HIVE?						
SIGNS OF ROBBERY AMONG THE BEES?						

HEALTH STATUS						
BEES SEEM WEAK OR LAZY?						
HIGH AMOUNT OF DEAD BEES?						
QUEEN BEE IS PRESENT / IDENTIFIABLE?						
INFESTATION BY ANTS / ANTS PRESENT?						
INFESTATION BY WAX MOTH / WAX MOTH PRESENT?						
NEGATIVE ODOUR NOTICEABLE?						

COLONY NAME	
DATE	
TIME	

WEATHER CONDITIONS						
Temperature	___	Sunny	Partly cloudy	Rain	Storm	Snow
Wind	___	☐	☐	☐	☐	☐

INSPECTION						
HIVE NUMBER	1	2	3	4	5	6

PRODUCTIVITY & REPRODUCTION						
AMOUNT OF HONEY						
GENERAL POPULATION						
AMOUNT OF BROOD						
AMOUNT OF SPACE						

BEHAVIOUR & ACTIVITIES						
USUAL ENTERING AND EXITING ACTIVITY?						
CALM BEHAVIOUR WHEN OPENING HIVE?						
BEES BRINGING POLLEN INTO HIVE?						
SIGNS OF ROBBERY AMONG THE BEES?						

HEALTH STATUS						
BEES SEEM WEAK OR LAZY?						
HIGH AMOUNT OF DEAD BEES?						
QUEEN BEE IS PRESENT / IDENTIFIABLE?						
INFESTATION BY ANTS / ANTS PRESENT?						
INFESTATION BY WAX MOTH / WAX MOTH PRESENT?						
NEGATIVE ODOUR NOTICEABLE?						

COLONY NAME	
DATE	
TIME	

WEATHER CONDITIONS						
	___					
	___	☐	☐	☐	☐	☐

INSPECTION						
HIVE NUMBER	1	2	3	4	5	6

PRODUCTIVITY & REPRODUCTION						
AMOUNT OF HONEY						
GENERAL POPULATION						
AMOUNT OF BROOD						
AMOUNT OF SPACE						

BEHAVIOUR & ACTIVITIES						
USUAL ENTERING AND EXITING ACTIVITY?						
CALM BEHAVIOUR WHEN OPENING HIVE?						
BEES BRINGING POLLEN INTO HIVE?						
SIGNS OF ROBBERY AMONG THE BEES?						

HEALTH STATUS						
BEES SEEM WEAK OR LAZY?						
HIGH AMOUNT OF DEAD BEES?						
QUEEN BEE IS PRESENT / IDENTIFIABLE?						
INFESTATION BY ANTS / ANTS PRESENT?						
INFESTATION BY WAX MOTH / WAX MOTH PRESENT?						
NEGATIVE ODOUR NOTICEABLE?						

COLONY NAME	
DATE	
TIME	

WEATHER CONDITIONS					
___	sunny	partly cloudy	rain	storm	snow
___	☐	☐	☐	☐	☐

INSPECTION						
HIVE NUMBER	1	2	3	4	5	6

PRODUCTIVITY & REPRODUCTION						
AMOUNT OF HONEY						
GENERAL POPULATION						
AMOUNT OF BROOD						
AMOUNT OF SPACE						

BEHAVIOUR & ACTIVITIES						
USUAL ENTERING AND EXITING ACTIVITY?						
CALM BEHAVIOUR WHEN OPENING HIVE?						
BEES BRINGING POLLEN INTO HIVE?						
SIGNS OF ROBBERY AMONG THE BEES?						

HEALTH STATUS						
BEES SEEM WEAK OR LAZY?						
HIGH AMOUNT OF DEAD BEES?						
QUEEN BEE IS PRESENT / IDENTIFIABLE?						
INFESTATION BY ANTS / ANTS PRESENT?						
INFESTATION BY WAX MOTH / WAX MOTH PRESENT?						
NEGATIVE ODOUR NOTICEABLE?						

COLONY NAME	
DATE	
TIME	

WEATHER CONDITIONS						
Temperature	___	☐	☐	☐	☐	☐
Wind	___	☐	☐	☐	☐	☐

INSPECTION						
HIVE NUMBER	1	2	3	4	5	6

PRODUCTIVITY & REPRODUCTION						
AMOUNT OF HONEY						
GENERAL POPULATION						
AMOUNT OF BROOD						
AMOUNT OF SPACE						

BEHAVIOUR & ACTIVITIES						
USUAL ENTERING AND EXITING ACTIVITY?						
CALM BEHAVIOUR WHEN OPENING HIVE?						
BEES BRINGING POLLEN INTO HIVE?						
SIGNS OF ROBBERY AMONG THE BEES?						

HEALTH STATUS						
BEES SEEM WEAK OR LAZY?						
HIGH AMOUNT OF DEAD BEES?						
QUEEN BEE IS PRESENT / IDENTIFIABLE?						
INFESTATION BY ANTS / ANTS PRESENT?						
INFESTATION BY WAX MOTH / WAX MOTH PRESENT?						
NEGATIVE ODOUR NOTICEABLE?						

COLONY NAME	
DATE	
TIME	

WEATHER CONDITIONS						
Temperature	___	☐	☐	☐	☐	☐
Wind	___					

INSPECTION						
HIVE NUMBER	1	2	3	4	5	6

PRODUCTIVITY & REPRODUCTION						
AMOUNT OF HONEY						
GENERAL POPULATION						
AMOUNT OF BROOD						
AMOUNT OF SPACE						

BEHAVIOUR & ACTIVITIES						
USUAL ENTERING AND EXITING ACTIVITY?						
CALM BEHAVIOUR WHEN OPENING HIVE?						
BEES BRINGING POLLEN INTO HIVE?						
SIGNS OF ROBBERY AMONG THE BEES?						

HEALTH STATUS						
BEES SEEM WEAK OR LAZY?						
HIGH AMOUNT OF DEAD BEES?						
QUEEN BEE IS PRESENT / IDENTIFIABLE?						
INFESTATION BY ANTS / ANTS PRESENT?						
INFESTATION BY WAX MOTH / WAX MOTH PRESENT?						
NEGATIVE ODOUR NOTICEABLE?						

COLONY NAME	
DATE	
TIME	

WEATHER CONDITIONS						
Temperature	___	Sunny	Partly cloudy	Rain	Storm	Snow
Wind	___	☐	☐	☐	☐	☐

INSPECTION						
HIVE NUMBER	1	2	3	4	5	6

PRODUCTIVITY & REPRODUCTION						
AMOUNT OF HONEY						
GENERAL POPULATION						
AMOUNT OF BROOD						
AMOUNT OF SPACE						

BEHAVIOUR & ACTIVITIES						
USUAL ENTERING AND EXITING ACTIVITY?						
CALM BEHAVIOUR WHEN OPENING HIVE?						
BEES BRINGING POLLEN INTO HIVE?						
SIGNS OF ROBBERY AMONG THE BEES?						

HEALTH STATUS						
BEES SEEM WEAK OR LAZY?						
HIGH AMOUNT OF DEAD BEES?						
QUEEN BEE IS PRESENT / IDENTIFIABLE?						
INFESTATION BY ANTS / ANTS PRESENT?						
INFESTATION BY WAX MOTH / WAX MOTH PRESENT?						
NEGATIVE ODOUR NOTICEABLE?						

COLONY NAME	
DATE	
TIME	

WEATHER CONDITIONS						
Temperature	___	☐	☐	☐	☐	☐
Wind	___					

INSPECTION						
HIVE NUMBER	1	2	3	4	5	6

PRODUCTIVITY & REPRODUCTION						
AMOUNT OF HONEY						
GENERAL POPULATION						
AMOUNT OF BROOD						
AMOUNT OF SPACE						

BEHAVIOUR & ACTIVITIES						
USUAL ENTERING AND EXITING ACTIVITY?						
CALM BEHAVIOUR WHEN OPENING HIVE?						
BEES BRINGING POLLEN INTO HIVE?						
SIGNS OF ROBBERY AMONG THE BEES?						

HEALTH STATUS						
BEES SEEM WEAK OR LAZY?						
HIGH AMOUNT OF DEAD BEES?						
QUEEN BEE IS PRESENT / IDENTIFIABLE?						
INFESTATION BY ANTS / ANTS PRESENT?						
INFESTATION BY WAX MOTH / WAX MOTH PRESENT?						
NEGATIVE ODOUR NOTICEABLE?						

COLONY NAME	
DATE	
TIME	

WEATHER CONDITIONS						
(temperature)	___	(sunny)	(partly cloudy)	(rain)	(storm)	(snow)
(wind)	___	☐	☐	☐	☐	☐

INSPECTION						
HIVE NUMBER	1	2	3	4	5	6

PRODUCTIVITY & REPRODUCTION						
AMOUNT OF HONEY						
GENERAL POPULATION						
AMOUNT OF BROOD						
AMOUNT OF SPACE						

BEHAVIOUR & ACTIVITIES						
USUAL ENTERING AND EXITING ACTIVITY?						
CALM BEHAVIOUR WHEN OPENING HIVE?						
BEES BRINGING POLLEN INTO HIVE?						
SIGNS OF ROBBERY AMONG THE BEES?						

HEALTH STATUS						
BEES SEEM WEAK OR LAZY?						
HIGH AMOUNT OF DEAD BEES?						
QUEEN BEE IS PRESENT / IDENTIFIABLE?						
INFESTATION BY ANTS / ANTS PRESENT?						
INFESTATION BY WAX MOTH / WAX MOTH PRESENT?						
NEGATIVE ODOUR NOTICEABLE?						

COLONY NAME	
DATE	
TIME	

WEATHER CONDITIONS						
Temperature	___	☐	☐	☐	☐	☐
Wind	___					

INSPECTION						
HIVE NUMBER	1	2	3	4	5	6

PRODUCTIVITY & REPRODUCTION						
AMOUNT OF HONEY						
GENERAL POPULATION						
AMOUNT OF BROOD						
AMOUNT OF SPACE						

BEHAVIOUR & ACTIVITIES						
USUAL ENTERING AND EXITING ACTIVITY?						
CALM BEHAVIOUR WHEN OPENING HIVE?						
BEES BRINGING POLLEN INTO HIVE?						
SIGNS OF ROBBERY AMONG THE BEES?						

HEALTH STATUS						
BEES SEEM WEAK OR LAZY?						
HIGH AMOUNT OF DEAD BEES?						
QUEEN BEE IS PRESENT / IDENTIFIABLE?						
INFESTATION BY ANTS / ANTS PRESENT?						
INFESTATION BY WAX MOTH / WAX MOTH PRESENT?						
NEGATIVE ODOUR NOTICEABLE?						

COLONY NAME	
DATE	
TIME	

WEATHER CONDITIONS						
Temperature	___	☐	☐	☐	☐	☐
Wind	___					

INSPECTION						
HIVE NUMBER	1	2	3	4	5	6

PRODUCTIVITY & REPRODUCTION						
AMOUNT OF HONEY						
GENERAL POPULATION						
AMOUNT OF BROOD						
AMOUNT OF SPACE						

BEHAVIOUR & ACTIVITIES						
USUAL ENTERING AND EXITING ACTIVITY?						
CALM BEHAVIOUR WHEN OPENING HIVE?						
BEES BRINGING POLLEN INTO HIVE?						
SIGNS OF ROBBERY AMONG THE BEES?						

HEALTH STATUS						
BEES SEEM WEAK OR LAZY?						
HIGH AMOUNT OF DEAD BEES?						
QUEEN BEE IS PRESENT / IDENTIFIABLE?						
INFESTATION BY ANTS / ANTS PRESENT?						
INFESTATION BY WAX MOTH / WAX MOTH PRESENT?						
NEGATIVE ODOUR NOTICEABLE?						

COLONY NAME	
DATE	
TIME	

WEATHER CONDITIONS						
Temperature	___	☐	☐	☐	☐	☐
Wind	___					

INSPECTION						
HIVE NUMBER	1	2	3	4	5	6

PRODUCTIVITY & REPRODUCTION						
AMOUNT OF HONEY						
GENERAL POPULATION						
AMOUNT OF BROOD						
AMOUNT OF SPACE						

BEHAVIOUR & ACTIVITIES						
USUAL ENTERING AND EXITING ACTIVITY?						
CALM BEHAVIOUR WHEN OPENING HIVE?						
BEES BRINGING POLLEN INTO HIVE?						
SIGNS OF ROBBERY AMONG THE BEES?						

HEALTH STATUS						
BEES SEEM WEAK OR LAZY?						
HIGH AMOUNT OF DEAD BEES?						
QUEEN BEE IS PRESENT / IDENTIFIABLE?						
INFESTATION BY ANTS / ANTS PRESENT?						
INFESTATION BY WAX MOTH / WAX MOTH PRESENT?						
NEGATIVE ODOUR NOTICEABLE?						

COLONY NAME	
DATE	
TIME	

WEATHER CONDITIONS						
Temperature	___	Sunny	Partly cloudy	Rain	Thunderstorm	Snow
Wind	___	☐	☐	☐	☐	☐

INSPECTION

HIVE NUMBER	1	2	3	4	5	6
PRODUCTIVITY & REPRODUCTION						
AMOUNT OF HONEY						
GENERAL POPULATION						
AMOUNT OF BROOD						
AMOUNT OF SPACE						
BEHAVIOUR & ACTIVITIES						
USUAL ENTERING AND EXITING ACTIVITY?						
CALM BEHAVIOUR WHEN OPENING HIVE?						
BEES BRINGING POLLEN INTO HIVE?						
SIGNS OF ROBBERY AMONG THE BEES?						
HEALTH STATUS						
BEES SEEM WEAK OR LAZY?						
HIGH AMOUNT OF DEAD BEES?						
QUEEN BEE IS PRESENT / IDENTIFIABLE?						
INFESTATION BY ANTS / ANTS PRESENT?						
INFESTATION BY WAX MOTH / WAX MOTH PRESENT?						
NEGATIVE ODOUR NOTICEABLE?						

COLONY NAME
DATE
TIME

WEATHER CONDITIONS						
	___					
	___	☐	☐	☐	☐	☐

INSPECTION

HIVE NUMBER	1	2	3	4	5	6

PRODUCTIVITY & REPRODUCTION

AMOUNT OF HONEY						
GENERAL POPULATION						
AMOUNT OF BROOD						
AMOUNT OF SPACE						

BEHAVIOUR & ACTIVITIES

USUAL ENTERING AND EXITING ACTIVITY?						
CALM BEHAVIOUR WHEN OPENING HIVE?						
BEES BRINGING POLLEN INTO HIVE?						
SIGNS OF ROBBERY AMONG THE BEES?						

HEALTH STATUS

BEES SEEM WEAK OR LAZY?						
HIGH AMOUNT OF DEAD BEES?						
QUEEN BEE IS PRESENT / IDENTIFIABLE?						
INFESTATION BY ANTS / ANTS PRESENT?						
INFESTATION BY WAX MOTH / WAX MOTH PRESENT?						
NEGATIVE ODOUR NOTICEABLE?						

COLONY NAME	
DATE	
TIME	

WEATHER CONDITIONS
___
___ ☐ ☐ ☐ ☐ ☐

INSPECTION						
HIVE NUMBER	1	2	3	4	5	6

PRODUCTIVITY & REPRODUCTION						
AMOUNT OF HONEY						
GENERAL POPULATION						
AMOUNT OF BROOD						
AMOUNT OF SPACE						

BEHAVIOUR & ACTIVITIES						
USUAL ENTERING AND EXITING ACTIVITY?						
CALM BEHAVIOUR WHEN OPENING HIVE?						
BEES BRINGING POLLEN INTO HIVE?						
SIGNS OF ROBBERY AMONG THE BEES?						

HEALTH STATUS						
BEES SEEM WEAK OR LAZY?						
HIGH AMOUNT OF DEAD BEES?						
QUEEN BEE IS PRESENT / IDENTIFIABLE?						
INFESTATION BY ANTS / ANTS PRESENT?						
INFESTATION BY WAX MOTH / WAX MOTH PRESENT?						
NEGATIVE ODOUR NOTICEABLE?						

COLONY NAME	
DATE	
TIME	

WEATHER CONDITIONS						
Temperature	___	☐	☐	☐	☐	☐
Wind	___					

INSPECTION						
HIVE NUMBER	1	2	3	4	5	6

PRODUCTIVITY & REPRODUCTION						
AMOUNT OF HONEY						
GENERAL POPULATION						
AMOUNT OF BROOD						
AMOUNT OF SPACE						

BEHAVIOUR & ACTIVITIES						
USUAL ENTERING AND EXITING ACTIVITY?						
CALM BEHAVIOUR WHEN OPENING HIVE?						
BEES BRINGING POLLEN INTO HIVE?						
SIGNS OF ROBBERY AMONG THE BEES?						

HEALTH STATUS						
BEES SEEM WEAK OR LAZY?						
HIGH AMOUNT OF DEAD BEES?						
QUEEN BEE IS PRESENT / IDENTIFIABLE?						
INFESTATION BY ANTS / ANTS PRESENT?						
INFESTATION BY WAX MOTH / WAX MOTH PRESENT?						
NEGATIVE ODOUR NOTICEABLE?						

COLONY NAME	
DATE	
TIME	

WEATHER CONDITIONS						
Temperature	___	☐	☐	☐	☐	☐
Wind	___					

INSPECTION						
HIVE NUMBER	1	2	3	4	5	6

PRODUCTIVITY & REPRODUCTION						
AMOUNT OF HONEY						
GENERAL POPULATION						
AMOUNT OF BROOD						
AMOUNT OF SPACE						

BEHAVIOUR & ACTIVITIES						
USUAL ENTERING AND EXITING ACTIVITY?						
CALM BEHAVIOUR WHEN OPENING HIVE?						
BEES BRINGING POLLEN INTO HIVE?						
SIGNS OF ROBBERY AMONG THE BEES?						

HEALTH STATUS						
BEES SEEM WEAK OR LAZY?						
HIGH AMOUNT OF DEAD BEES?						
QUEEN BEE IS PRESENT / IDENTIFIABLE?						
INFESTATION BY ANTS / ANTS PRESENT?						
INFESTATION BY WAX MOTH / WAX MOTH PRESENT?						
NEGATIVE ODOUR NOTICEABLE?						

COLONY NAME	
DATE	
TIME	

WEATHER CONDITIONS					
___					
___	☐	☐	☐	☐	☐

INSPECTION						
HIVE NUMBER	1	2	3	4	5	6

PRODUCTIVITY & REPRODUCTION						
AMOUNT OF HONEY						
GENERAL POPULATION						
AMOUNT OF BROOD						
AMOUNT OF SPACE						

BEHAVIOUR & ACTIVITIES						
USUAL ENTERING AND EXITING ACTIVITY?						
CALM BEHAVIOUR WHEN OPENING HIVE?						
BEES BRINGING POLLEN INTO HIVE?						
SIGNS OF ROBBERY AMONG THE BEES?						

HEALTH STATUS						
BEES SEEM WEAK OR LAZY?						
HIGH AMOUNT OF DEAD BEES?						
QUEEN BEE IS PRESENT / IDENTIFIABLE?						
INFESTATION BY ANTS / ANTS PRESENT?						
INFESTATION BY WAX MOTH / WAX MOTH PRESENT?						
NEGATIVE ODOUR NOTICEABLE?						

COLONY NAME	
DATE	
TIME	

WEATHER CONDITIONS						
Temperature	___	Sunny	Partly cloudy	Rain	Storm	Snow
Wind	___	☐	☐	☐	☐	☐

INSPECTION						
HIVE NUMBER	1	2	3	4	5	6

PRODUCTIVITY & REPRODUCTION						
AMOUNT OF HONEY						
GENERAL POPULATION						
AMOUNT OF BROOD						
AMOUNT OF SPACE						

BEHAVIOUR & ACTIVITIES						
USUAL ENTERING AND EXITING ACTIVITY?						
CALM BEHAVIOUR WHEN OPENING HIVE?						
BEES BRINGING POLLEN INTO HIVE?						
SIGNS OF ROBBERY AMONG THE BEES?						

HEALTH STATUS						
BEES SEEM WEAK OR LAZY?						
HIGH AMOUNT OF DEAD BEES?						
QUEEN BEE IS PRESENT / IDENTIFIABLE?						
INFESTATION BY ANTS / ANTS PRESENT?						
INFESTATION BY WAX MOTH / WAX MOTH PRESENT?						
NEGATIVE ODOUR NOTICEABLE?						

COLONY NAME	
DATE	
TIME	

WEATHER CONDITIONS						
Temperature	___	☐	☐	☐	☐	☐
Wind	___	☐	☐	☐	☐	☐

INSPECTION						
HIVE NUMBER	1	2	3	4	5	6

PRODUCTIVITY & REPRODUCTION						
AMOUNT OF HONEY						
GENERAL POPULATION						
AMOUNT OF BROOD						
AMOUNT OF SPACE						

BEHAVIOUR & ACTIVITIES						
USUAL ENTERING AND EXITING ACTIVITY?						
CALM BEHAVIOUR WHEN OPENING HIVE?						
BEES BRINGING POLLEN INTO HIVE?						
SIGNS OF ROBBERY AMONG THE BEES?						

HEALTH STATUS						
BEES SEEM WEAK OR LAZY?						
HIGH AMOUNT OF DEAD BEES?						
QUEEN BEE IS PRESENT / IDENTIFIABLE?						
INFESTATION BY ANTS / ANTS PRESENT?						
INFESTATION BY WAX MOTH / WAX MOTH PRESENT?						
NEGATIVE ODOUR NOTICEABLE?						

COLONY NAME	
DATE	
TIME	

WEATHER CONDITIONS						
	___					
	___	☐	☐	☐	☐	☐

INSPECTION						
HIVE NUMBER	1	2	3	4	5	6

PRODUCTIVITY & REPRODUCTION						
AMOUNT OF HONEY						
GENERAL POPULATION						
AMOUNT OF BROOD						
AMOUNT OF SPACE						

BEHAVIOUR & ACTIVITIES						
USUAL ENTERING AND EXITING ACTIVITY?						
CALM BEHAVIOUR WHEN OPENING HIVE?						
BEES BRINGING POLLEN INTO HIVE?						
SIGNS OF ROBBERY AMONG THE BEES?						

HEALTH STATUS						
BEES SEEM WEAK OR LAZY?						
HIGH AMOUNT OF DEAD BEES?						
QUEEN BEE IS PRESENT / IDENTIFIABLE?						
INFESTATION BY ANTS / ANTS PRESENT?						
INFESTATION BY WAX MOTH / WAX MOTH PRESENT?						
NEGATIVE ODOUR NOTICEABLE?						

COLONY NAME
DATE
TIME

WEATHER CONDITIONS						
Temperature	___	Sunny	Partly cloudy	Rain	Storm	Snow
Wind	___	☐	☐	☐	☐	☐

INSPECTION						
HIVE NUMBER	1	2	3	4	5	6

PRODUCTIVITY & REPRODUCTION						
AMOUNT OF HONEY						
GENERAL POPULATION						
AMOUNT OF BROOD						
AMOUNT OF SPACE						

BEHAVIOUR & ACTIVITIES						
USUAL ENTERING AND EXITING ACTIVITY?						
CALM BEHAVIOUR WHEN OPENING HIVE?						
BEES BRINGING POLLEN INTO HIVE?						
SIGNS OF ROBBERY AMONG THE BEES?						

HEALTH STATUS						
BEES SEEM WEAK OR LAZY?						
HIGH AMOUNT OF DEAD BEES?						
QUEEN BEE IS PRESENT / IDENTIFIABLE?						
INFESTATION BY ANTS / ANTS PRESENT?						
INFESTATION BY WAX MOTH / WAX MOTH PRESENT?						
NEGATIVE ODOUR NOTICEABLE?						

COLONY NAME	
DATE	
TIME	

WEATHER CONDITIONS						
Temperature	___	Sunny	Partly cloudy	Rain	Storm	Snow
Wind	___	☐	☐	☐	☐	☐

INSPECTION						
HIVE NUMBER	1	2	3	4	5	6

PRODUCTIVITY & REPRODUCTION						
AMOUNT OF HONEY						
GENERAL POPULATION						
AMOUNT OF BROOD						
AMOUNT OF SPACE						

BEHAVIOUR & ACTIVITIES						
USUAL ENTERING AND EXITING ACTIVITY?						
CALM BEHAVIOUR WHEN OPENING HIVE?						
BEES BRINGING POLLEN INTO HIVE?						
SIGNS OF ROBBERY AMONG THE BEES?						

HEALTH STATUS						
BEES SEEM WEAK OR LAZY?						
HIGH AMOUNT OF DEAD BEES?						
QUEEN BEE IS PRESENT / IDENTIFIABLE?						
INFESTATION BY ANTS / ANTS PRESENT?						
INFESTATION BY WAX MOTH / WAX MOTH PRESENT?						
NEGATIVE ODOUR NOTICEABLE?						

COLONY NAME	
DATE	
TIME	

WEATHER CONDITIONS						
Temperature	___	Sunny	Partly cloudy	Rain	Storm	Snow
Wind	___	☐	☐	☐	☐	☐

INSPECTION						
HIVE NUMBER	1	2	3	4	5	6

PRODUCTIVITY & REPRODUCTION						
AMOUNT OF HONEY						
GENERAL POPULATION						
AMOUNT OF BROOD						
AMOUNT OF SPACE						

BEHAVIOUR & ACTIVITIES						
USUAL ENTERING AND EXITING ACTIVITY?						
CALM BEHAVIOUR WHEN OPENING HIVE?						
BEES BRINGING POLLEN INTO HIVE?						
SIGNS OF ROBBERY AMONG THE BEES?						

HEALTH STATUS						
BEES SEEM WEAK OR LAZY?						
HIGH AMOUNT OF DEAD BEES?						
QUEEN BEE IS PRESENT / IDENTIFIABLE?						
INFESTATION BY ANTS / ANTS PRESENT?						
INFESTATION BY WAX MOTH / WAX MOTH PRESENT?						
NEGATIVE ODOUR NOTICEABLE?						

COLONY NAME	
DATE	
TIME	

WEATHER CONDITIONS						
(temperature)	___	☐	☐	☐	☐	☐
(wind)	___					

INSPECTION						
HIVE NUMBER	1	2	3	4	5	6

PRODUCTIVITY & REPRODUCTION						
AMOUNT OF HONEY						
GENERAL POPULATION						
AMOUNT OF BROOD						
AMOUNT OF SPACE						

BEHAVIOUR & ACTIVITIES						
USUAL ENTERING AND EXITING ACTIVITY?						
CALM BEHAVIOUR WHEN OPENING HIVE?						
BEES BRINGING POLLEN INTO HIVE?						
SIGNS OF ROBBERY AMONG THE BEES?						

HEALTH STATUS						
BEES SEEM WEAK OR LAZY?						
HIGH AMOUNT OF DEAD BEES?						
QUEEN BEE IS PRESENT / IDENTIFIABLE?						
INFESTATION BY ANTS / ANTS PRESENT?						
INFESTATION BY WAX MOTH / WAX MOTH PRESENT?						
NEGATIVE ODOUR NOTICEABLE?						

COLONY NAME	
DATE	
TIME	

WEATHER CONDITIONS					
___	☼	partly cloudy	rain	storm	snow
___	☐	☐	☐	☐	☐

INSPECTION						
HIVE NUMBER	1	2	3	4	5	6

PRODUCTIVITY & REPRODUCTION						
AMOUNT OF HONEY						
GENERAL POPULATION						
AMOUNT OF BROOD						
AMOUNT OF SPACE						

BEHAVIOUR & ACTIVITIES						
USUAL ENTERING AND EXITING ACTIVITY?						
CALM BEHAVIOUR WHEN OPENING HIVE?						
BEES BRINGING POLLEN INTO HIVE?						
SIGNS OF ROBBERY AMONG THE BEES?						

HEALTH STATUS						
BEES SEEM WEAK OR LAZY?						
HIGH AMOUNT OF DEAD BEES?						
QUEEN BEE IS PRESENT / IDENTIFIABLE?						
INFESTATION BY ANTS / ANTS PRESENT?						
INFESTATION BY WAX MOTH / WAX MOTH PRESENT?						
NEGATIVE ODOUR NOTICEABLE?						

COLONY NAME	
DATE	
TIME	

WEATHER CONDITIONS						
Temperature	___	Sunny	Partly cloudy	Rain	Storm	Snow
Wind	___	☐	☐	☐	☐	☐

INSPECTION						
HIVE NUMBER	1	2	3	4	5	6

PRODUCTIVITY & REPRODUCTION						
AMOUNT OF HONEY						
GENERAL POPULATION						
AMOUNT OF BROOD						
AMOUNT OF SPACE						

BEHAVIOUR & ACTIVITIES						
USUAL ENTERING AND EXITING ACTIVITY?						
CALM BEHAVIOUR WHEN OPENING HIVE?						
BEES BRINGING POLLEN INTO HIVE?						
SIGNS OF ROBBERY AMONG THE BEES?						

HEALTH STATUS						
BEES SEEM WEAK OR LAZY?						
HIGH AMOUNT OF DEAD BEES?						
QUEEN BEE IS PRESENT / IDENTIFIABLE?						
INFESTATION BY ANTS / ANTS PRESENT?						
INFESTATION BY WAX MOTH / WAX MOTH PRESENT?						
NEGATIVE ODOUR NOTICEABLE?						

COLONY NAME	
DATE	
TIME	

WEATHER CONDITIONS						
Temperature	___	☐	☐	☐	☐	☐
Wind	___					

INSPECTION						
HIVE NUMBER	1	2	3	4	5	6

PRODUCTIVITY & REPRODUCTION						
AMOUNT OF HONEY						
GENERAL POPULATION						
AMOUNT OF BROOD						
AMOUNT OF SPACE						

BEHAVIOUR & ACTIVITIES						
USUAL ENTERING AND EXITING ACTIVITY?						
CALM BEHAVIOUR WHEN OPENING HIVE?						
BEES BRINGING POLLEN INTO HIVE?						
SIGNS OF ROBBERY AMONG THE BEES?						

HEALTH STATUS						
BEES SEEM WEAK OR LAZY?						
HIGH AMOUNT OF DEAD BEES?						
QUEEN BEE IS PRESENT / IDENTIFIABLE?						
INFESTATION BY ANTS / ANTS PRESENT?						
INFESTATION BY WAX MOTH / WAX MOTH PRESENT?						
NEGATIVE ODOUR NOTICEABLE?						

COLONY NAME	
DATE	
TIME	

WEATHER CONDITIONS						
Temperature	___	Sunny	Partly cloudy	Rain	Storm	Snow
Wind	___	☐	☐	☐	☐	☐

INSPECTION						
HIVE NUMBER	1	2	3	4	5	6

PRODUCTIVITY & REPRODUCTION						
AMOUNT OF HONEY						
GENERAL POPULATION						
AMOUNT OF BROOD						
AMOUNT OF SPACE						

BEHAVIOUR & ACTIVITIES						
USUAL ENTERING AND EXITING ACTIVITY?						
CALM BEHAVIOUR WHEN OPENING HIVE?						
BEES BRINGING POLLEN INTO HIVE?						
SIGNS OF ROBBERY AMONG THE BEES?						

HEALTH STATUS						
BEES SEEM WEAK OR LAZY?						
HIGH AMOUNT OF DEAD BEES?						
QUEEN BEE IS PRESENT / IDENTIFIABLE?						
INFESTATION BY ANTS / ANTS PRESENT?						
INFESTATION BY WAX MOTH / WAX MOTH PRESENT?						
NEGATIVE ODOUR NOTICEABLE?						

COLONY NAME	
DATE	
TIME	

WEATHER CONDITIONS						
Temperature	___	☐	☐	☐	☐	☐
Wind	___					

INSPECTION						
HIVE NUMBER	1	2	3	4	5	6

PRODUCTIVITY & REPRODUCTION						
AMOUNT OF HONEY						
GENERAL POPULATION						
AMOUNT OF BROOD						
AMOUNT OF SPACE						

BEHAVIOUR & ACTIVITIES						
USUAL ENTERING AND EXITING ACTIVITY?						
CALM BEHAVIOUR WHEN OPENING HIVE?						
BEES BRINGING POLLEN INTO HIVE?						
SIGNS OF ROBBERY AMONG THE BEES?						

HEALTH STATUS						
BEES SEEM WEAK OR LAZY?						
HIGH AMOUNT OF DEAD BEES?						
QUEEN BEE IS PRESENT / IDENTIFIABLE?						
INFESTATION BY ANTS / ANTS PRESENT?						
INFESTATION BY WAX MOTH / WAX MOTH PRESENT?						
NEGATIVE ODOUR NOTICEABLE?						

COLONY NAME	
DATE	
TIME	

WEATHER CONDITIONS					
___					
___	☐	☐	☐	☐	☐

INSPECTION						
HIVE NUMBER	1	2	3	4	5	6

PRODUCTIVITY & REPRODUCTION						
AMOUNT OF HONEY						
GENERAL POPULATION						
AMOUNT OF BROOD						
AMOUNT OF SPACE						

BEHAVIOUR & ACTIVITIES						
USUAL ENTERING AND EXITING ACTIVITY?						
CALM BEHAVIOUR WHEN OPENING HIVE?						
BEES BRINGING POLLEN INTO HIVE?						
SIGNS OF ROBBERY AMONG THE BEES?						

HEALTH STATUS						
BEES SEEM WEAK OR LAZY?						
HIGH AMOUNT OF DEAD BEES?						
QUEEN BEE IS PRESENT / IDENTIFIABLE?						
INFESTATION BY ANTS / ANTS PRESENT?						
INFESTATION BY WAX MOTH / WAX MOTH PRESENT?						
NEGATIVE ODOUR NOTICEABLE?						

COLONY NAME	
DATE	
TIME	

WEATHER CONDITIONS						
Temperature	____	☐	☐	☐	☐	☐
Wind	____					

INSPECTION

HIVE NUMBER	1	2	3	4	5	6

PRODUCTIVITY & REPRODUCTION

AMOUNT OF HONEY						
GENERAL POPULATION						
AMOUNT OF BROOD						
AMOUNT OF SPACE						

BEHAVIOUR & ACTIVITIES

USUAL ENTERING AND EXITING ACTIVITY?						
CALM BEHAVIOUR WHEN OPENING HIVE?						
BEES BRINGING POLLEN INTO HIVE?						
SIGNS OF ROBBERY AMONG THE BEES?						

HEALTH STATUS

BEES SEEM WEAK OR LAZY?						
HIGH AMOUNT OF DEAD BEES?						
QUEEN BEE IS PRESENT / IDENTIFIABLE?						
INFESTATION BY ANTS / ANTS PRESENT?						
INFESTATION BY WAX MOTH / WAX MOTH PRESENT?						
NEGATIVE ODOUR NOTICEABLE?						

COLONY NAME	
DATE	
TIME	

WEATHER CONDITIONS						
Temperature	___	☐	☐	☐	☐	☐
Wind	___	☐	☐	☐	☐	☐

INSPECTION						
HIVE NUMBER	1	2	3	4	5	6

PRODUCTIVITY & REPRODUCTION						
AMOUNT OF HONEY						
GENERAL POPULATION						
AMOUNT OF BROOD						
AMOUNT OF SPACE						

BEHAVIOUR & ACTIVITIES						
USUAL ENTERING AND EXITING ACTIVITY?						
CALM BEHAVIOUR WHEN OPENING HIVE?						
BEES BRINGING POLLEN INTO HIVE?						
SIGNS OF ROBBERY AMONG THE BEES?						

HEALTH STATUS						
BEES SEEM WEAK OR LAZY?						
HIGH AMOUNT OF DEAD BEES?						
QUEEN BEE IS PRESENT / IDENTIFIABLE?						
INFESTATION BY ANTS / ANTS PRESENT?						
INFESTATION BY WAX MOTH / WAX MOTH PRESENT?						
NEGATIVE ODOUR NOTICEABLE?						

COLONY NAME	
DATE	
TIME	

WEATHER CONDITIONS						
Temperature	___	☐	☐	☐	☐	☐
Wind	___	☐	☐	☐	☐	☐

INSPECTION

HIVE NUMBER	1	2	3	4	5	6

PRODUCTIVITY & REPRODUCTION

AMOUNT OF HONEY						
GENERAL POPULATION						
AMOUNT OF BROOD						
AMOUNT OF SPACE						

BEHAVIOUR & ACTIVITIES

USUAL ENTERING AND EXITING ACTIVITY?						
CALM BEHAVIOUR WHEN OPENING HIVE?						
BEES BRINGING POLLEN INTO HIVE?						
SIGNS OF ROBBERY AMONG THE BEES?						

HEALTH STATUS

BEES SEEM WEAK OR LAZY?						
HIGH AMOUNT OF DEAD BEES?						
QUEEN BEE IS PRESENT / IDENTIFIABLE?						
INFESTATION BY ANTS / ANTS PRESENT?						
INFESTATION BY WAX MOTH / WAX MOTH PRESENT?						
NEGATIVE ODOUR NOTICEABLE?						

COLONY NAME
DATE
TIME

WEATHER CONDITIONS						
Temperature	___	☐	☐	☐	☐	☐
Wind	___					

INSPECTION						
HIVE NUMBER	1	2	3	4	5	6

PRODUCTIVITY & REPRODUCTION						
AMOUNT OF HONEY						
GENERAL POPULATION						
AMOUNT OF BROOD						
AMOUNT OF SPACE						

BEHAVIOUR & ACTIVITIES						
USUAL ENTERING AND EXITING ACTIVITY?						
CALM BEHAVIOUR WHEN OPENING HIVE?						
BEES BRINGING POLLEN INTO HIVE?						
SIGNS OF ROBBERY AMONG THE BEES?						

HEALTH STATUS						
BEES SEEM WEAK OR LAZY?						
HIGH AMOUNT OF DEAD BEES?						
QUEEN BEE IS PRESENT / IDENTIFIABLE?						
INFESTATION BY ANTS / ANTS PRESENT?						
INFESTATION BY WAX MOTH / WAX MOTH PRESENT?						
NEGATIVE ODOUR NOTICEABLE?						

COLONY NAME	
DATE	
TIME	

WEATHER CONDITIONS						
Temperature	___	☐	☐	☐	☐	☐
Wind	___	☐	☐	☐	☐	☐

INSPECTION						
HIVE NUMBER	1	2	3	4	5	6

PRODUCTIVITY & REPRODUCTION						
AMOUNT OF HONEY						
GENERAL POPULATION						
AMOUNT OF BROOD						
AMOUNT OF SPACE						

BEHAVIOUR & ACTIVITIES						
USUAL ENTERING AND EXITING ACTIVITY?						
CALM BEHAVIOUR WHEN OPENING HIVE?						
BEES BRINGING POLLEN INTO HIVE?						
SIGNS OF ROBBERY AMONG THE BEES?						

HEALTH STATUS						
BEES SEEM WEAK OR LAZY?						
HIGH AMOUNT OF DEAD BEES?						
QUEEN BEE IS PRESENT / IDENTIFIABLE?						
INFESTATION BY ANTS / ANTS PRESENT?						
INFESTATION BY WAX MOTH / WAX MOTH PRESENT?						
NEGATIVE ODOUR NOTICEABLE?						

COLONY NAME	
DATE	
TIME	

WEATHER CONDITIONS						
Temperature	___	☐	☐	☐	☐	☐
Wind	___	☐	☐	☐	☐	☐

INSPECTION

HIVE NUMBER	1	2	3	4	5	6

PRODUCTIVITY & REPRODUCTION

AMOUNT OF HONEY						
GENERAL POPULATION						
AMOUNT OF BROOD						
AMOUNT OF SPACE						

BEHAVIOUR & ACTIVITIES

USUAL ENTERING AND EXITING ACTIVITY?						
CALM BEHAVIOUR WHEN OPENING HIVE?						
BEES BRINGING POLLEN INTO HIVE?						
SIGNS OF ROBBERY AMONG THE BEES?						

HEALTH STATUS

BEES SEEM WEAK OR LAZY?						
HIGH AMOUNT OF DEAD BEES?						
QUEEN BEE IS PRESENT / IDENTIFIABLE?						
INFESTATION BY ANTS / ANTS PRESENT?						
INFESTATION BY WAX MOTH / WAX MOTH PRESENT?						
NEGATIVE ODOUR NOTICEABLE?						

COLONY NAME	
DATE	
TIME	

WEATHER CONDITIONS						
	___					
	___	☐	☐	☐	☐	☐

INSPECTION

HIVE NUMBER	1	2	3	4	5	6

PRODUCTIVITY & REPRODUCTION

AMOUNT OF HONEY						
GENERAL POPULATION						
AMOUNT OF BROOD						
AMOUNT OF SPACE						

BEHAVIOUR & ACTIVITIES

USUAL ENTERING AND EXITING ACTIVITY?						
CALM BEHAVIOUR WHEN OPENING HIVE?						
BEES BRINGING POLLEN INTO HIVE?						
SIGNS OF ROBBERY AMONG THE BEES?						

HEALTH STATUS

BEES SEEM WEAK OR LAZY?						
HIGH AMOUNT OF DEAD BEES?						
QUEEN BEE IS PRESENT / IDENTIFIABLE?						
INFESTATION BY ANTS / ANTS PRESENT?						
INFESTATION BY WAX MOTH / WAX MOTH PRESENT?						
NEGATIVE ODOUR NOTICEABLE?						

COLONY NAME	
DATE	
TIME	

WEATHER CONDITIONS						
Temperature	___	☐	☐	☐	☐	☐
Wind	___	☐	☐	☐	☐	☐

INSPECTION						
HIVE NUMBER	1	2	3	4	5	6

PRODUCTIVITY & REPRODUCTION						
AMOUNT OF HONEY						
GENERAL POPULATION						
AMOUNT OF BROOD						
AMOUNT OF SPACE						

BEHAVIOUR & ACTIVITIES						
USUAL ENTERING AND EXITING ACTIVITY?						
CALM BEHAVIOUR WHEN OPENING HIVE?						
BEES BRINGING POLLEN INTO HIVE?						
SIGNS OF ROBBERY AMONG THE BEES?						

HEALTH STATUS						
BEES SEEM WEAK OR LAZY?						
HIGH AMOUNT OF DEAD BEES?						
QUEEN BEE IS PRESENT / IDENTIFIABLE?						
INFESTATION BY ANTS / ANTS PRESENT?						
INFESTATION BY WAX MOTH / WAX MOTH PRESENT?						
NEGATIVE ODOUR NOTICEABLE?						

COLONY NAME	
DATE	
TIME	

WEATHER CONDITIONS						
Temperature	___	☐	☐	☐	☐	☐
Wind	___					

INSPECTION						
HIVE NUMBER	1	2	3	4	5	6

PRODUCTIVITY & REPRODUCTION						
AMOUNT OF HONEY						
GENERAL POPULATION						
AMOUNT OF BROOD						
AMOUNT OF SPACE						

BEHAVIOUR & ACTIVITIES						
USUAL ENTERING AND EXITING ACTIVITY?						
CALM BEHAVIOUR WHEN OPENING HIVE?						
BEES BRINGING POLLEN INTO HIVE?						
SIGNS OF ROBBERY AMONG THE BEES?						

HEALTH STATUS						
BEES SEEM WEAK OR LAZY?						
HIGH AMOUNT OF DEAD BEES?						
QUEEN BEE IS PRESENT / IDENTIFIABLE?						
INFESTATION BY ANTS / ANTS PRESENT?						
INFESTATION BY WAX MOTH / WAX MOTH PRESENT?						
NEGATIVE ODOUR NOTICEABLE?						

COLONY NAME	
DATE	
TIME	

WEATHER CONDITIONS						
Temperature	___	Sunny	Partly cloudy	Rain	Storm	Snow
Wind	___	☐	☐	☐	☐	☐

INSPECTION						
HIVE NUMBER	1	2	3	4	5	6

PRODUCTIVITY & REPRODUCTION						
AMOUNT OF HONEY						
GENERAL POPULATION						
AMOUNT OF BROOD						
AMOUNT OF SPACE						

BEHAVIOUR & ACTIVITIES						
USUAL ENTERING AND EXITING ACTIVITY?						
CALM BEHAVIOUR WHEN OPENING HIVE?						
BEES BRINGING POLLEN INTO HIVE?						
SIGNS OF ROBBERY AMONG THE BEES?						

HEALTH STATUS						
BEES SEEM WEAK OR LAZY?						
HIGH AMOUNT OF DEAD BEES?						
QUEEN BEE IS PRESENT / IDENTIFIABLE?						
INFESTATION BY ANTS / ANTS PRESENT?						
INFESTATION BY WAX MOTH / WAX MOTH PRESENT?						
NEGATIVE ODOUR NOTICEABLE?						

COLONY NAME	
DATE	
TIME	

WEATHER CONDITIONS						
Temperature	___	Sunny ☐	Partly cloudy ☐	Rain ☐	Storm ☐	Snow ☐
Wind	___					

INSPECTION						
HIVE NUMBER	1	2	3	4	5	6

PRODUCTIVITY & REPRODUCTION						
AMOUNT OF HONEY						
GENERAL POPULATION						
AMOUNT OF BROOD						
AMOUNT OF SPACE						

BEHAVIOUR & ACTIVITIES						
USUAL ENTERING AND EXITING ACTIVITY?						
CALM BEHAVIOUR WHEN OPENING HIVE?						
BEES BRINGING POLLEN INTO HIVE?						
SIGNS OF ROBBERY AMONG THE BEES?						

HEALTH STATUS						
BEES SEEM WEAK OR LAZY?						
HIGH AMOUNT OF DEAD BEES?						
QUEEN BEE IS PRESENT / IDENTIFIABLE?						
INFESTATION BY ANTS / ANTS PRESENT?						
INFESTATION BY WAX MOTH / WAX MOTH PRESENT?						
NEGATIVE ODOUR NOTICEABLE?						

COLONY NAME	
DATE	
TIME	

WEATHER CONDITIONS						
Temperature	___	☐	☐	☐	☐	☐
Wind	___					

INSPECTION						
HIVE NUMBER	1	2	3	4	5	6

PRODUCTIVITY & REPRODUCTION						
AMOUNT OF HONEY						
GENERAL POPULATION						
AMOUNT OF BROOD						
AMOUNT OF SPACE						

BEHAVIOUR & ACTIVITIES						
USUAL ENTERING AND EXITING ACTIVITY?						
CALM BEHAVIOUR WHEN OPENING HIVE?						
BEES BRINGING POLLEN INTO HIVE?						
SIGNS OF ROBBERY AMONG THE BEES?						

HEALTH STATUS						
BEES SEEM WEAK OR LAZY?						
HIGH AMOUNT OF DEAD BEES?						
QUEEN BEE IS PRESENT / IDENTIFIABLE?						
INFESTATION BY ANTS / ANTS PRESENT?						
INFESTATION BY WAX MOTH / WAX MOTH PRESENT?						
NEGATIVE ODOUR NOTICEABLE?						

COLONY NAME
DATE
TIME

WEATHER CONDITIONS
___
___

INSPECTION						
HIVE NUMBER	1	2	3	4	5	6

PRODUCTIVITY & REPRODUCTION						
AMOUNT OF HONEY						
GENERAL POPULATION						
AMOUNT OF BROOD						
AMOUNT OF SPACE						

BEHAVIOUR & ACTIVITIES						
USUAL ENTERING AND EXITING ACTIVITY?						
CALM BEHAVIOUR WHEN OPENING HIVE?						
BEES BRINGING POLLEN INTO HIVE?						
SIGNS OF ROBBERY AMONG THE BEES?						

HEALTH STATUS						
BEES SEEM WEAK OR LAZY?						
HIGH AMOUNT OF DEAD BEES?						
QUEEN BEE IS PRESENT / IDENTIFIABLE?						
INFESTATION BY ANTS / ANTS PRESENT?						
INFESTATION BY WAX MOTH / WAX MOTH PRESENT?						
NEGATIVE ODOUR NOTICEABLE?						

COLONY NAME
DATE
TIME

WEATHER CONDITIONS						
	___					
	___	☐	☐	☐	☐	☐

INSPECTION						
HIVE NUMBER	1	2	3	4	5	6

PRODUCTIVITY & REPRODUCTION						
AMOUNT OF HONEY						
GENERAL POPULATION						
AMOUNT OF BROOD						
AMOUNT OF SPACE						

BEHAVIOUR & ACTIVITIES						
USUAL ENTERING AND EXITING ACTIVITY?						
CALM BEHAVIOUR WHEN OPENING HIVE?						
BEES BRINGING POLLEN INTO HIVE?						
SIGNS OF ROBBERY AMONG THE BEES?						

HEALTH STATUS						
BEES SEEM WEAK OR LAZY?						
HIGH AMOUNT OF DEAD BEES?						
QUEEN BEE IS PRESENT / IDENTIFIABLE?						
INFESTATION BY ANTS / ANTS PRESENT?						
INFESTATION BY WAX MOTH / WAX MOTH PRESENT?						
NEGATIVE ODOUR NOTICEABLE?						

COLONY NAME	
DATE	
TIME	

WEATHER CONDITIONS						
Temperature	___	☐	☐	☐	☐	☐
Wind	___					

INSPECTION						
HIVE NUMBER	1	2	3	4	5	6

PRODUCTIVITY & REPRODUCTION						
AMOUNT OF HONEY						
GENERAL POPULATION						
AMOUNT OF BROOD						
AMOUNT OF SPACE						

BEHAVIOUR & ACTIVITIES						
USUAL ENTERING AND EXITING ACTIVITY?						
CALM BEHAVIOUR WHEN OPENING HIVE?						
BEES BRINGING POLLEN INTO HIVE?						
SIGNS OF ROBBERY AMONG THE BEES?						

HEALTH STATUS						
BEES SEEM WEAK OR LAZY?						
HIGH AMOUNT OF DEAD BEES?						
QUEEN BEE IS PRESENT / IDENTIFIABLE?						
INFESTATION BY ANTS / ANTS PRESENT?						
INFESTATION BY WAX MOTH / WAX MOTH PRESENT?						
NEGATIVE ODOUR NOTICEABLE?						

COLONY NAME	
DATE	
TIME	

WEATHER CONDITIONS						
	___					
	___	☐	☐	☐	☐	☐

INSPECTION

HIVE NUMBER	1	2	3	4	5	6

PRODUCTIVITY & REPRODUCTION

AMOUNT OF HONEY						
GENERAL POPULATION						
AMOUNT OF BROOD						
AMOUNT OF SPACE						

BEHAVIOUR & ACTIVITIES

USUAL ENTERING AND EXITING ACTIVITY?						
CALM BEHAVIOUR WHEN OPENING HIVE?						
BEES BRINGING POLLEN INTO HIVE?						
SIGNS OF ROBBERY AMONG THE BEES?						

HEALTH STATUS

BEES SEEM WEAK OR LAZY?						
HIGH AMOUNT OF DEAD BEES?						
QUEEN BEE IS PRESENT / IDENTIFIABLE?						
INFESTATION BY ANTS / ANTS PRESENT?						
INFESTATION BY WAX MOTH / WAX MOTH PRESENT?						
NEGATIVE ODOUR NOTICEABLE?						

COLONY NAME	
DATE	
TIME	

WEATHER CONDITIONS						
Temperature	____	Sunny	Partly cloudy	Rain	Storm	Snow
Wind	____	☐	☐	☐	☐	☐

INSPECTION

HIVE NUMBER	1	2	3	4	5	6

PRODUCTIVITY & REPRODUCTION

	1	2	3	4	5	6
AMOUNT OF HONEY						
GENERAL POPULATION						
AMOUNT OF BROOD						
AMOUNT OF SPACE						

BEHAVIOUR & ACTIVITIES

	1	2	3	4	5	6
USUAL ENTERING AND EXITING ACTIVITY?						
CALM BEHAVIOUR WHEN OPENING HIVE?						
BEES BRINGING POLLEN INTO HIVE?						
SIGNS OF ROBBERY AMONG THE BEES?						

HEALTH STATUS

	1	2	3	4	5	6
BEES SEEM WEAK OR LAZY?						
HIGH AMOUNT OF DEAD BEES?						
QUEEN BEE IS PRESENT / IDENTIFIABLE?						
INFESTATION BY ANTS / ANTS PRESENT?						
INFESTATION BY WAX MOTH / WAX MOTH PRESENT?						
NEGATIVE ODOUR NOTICEABLE?						

COLONY NAME	
DATE	
TIME	

WEATHER CONDITIONS						
	___	☐	☐	☐	☐	☐
	___					

INSPECTION						
HIVE NUMBER	1	2	3	4	5	6

PRODUCTIVITY & REPRODUCTION						
AMOUNT OF HONEY						
GENERAL POPULATION						
AMOUNT OF BROOD						
AMOUNT OF SPACE						

BEHAVIOUR & ACTIVITIES						
USUAL ENTERING AND EXITING ACTIVITY?						
CALM BEHAVIOUR WHEN OPENING HIVE?						
BEES BRINGING POLLEN INTO HIVE?						
SIGNS OF ROBBERY AMONG THE BEES?						

HEALTH STATUS						
BEES SEEM WEAK OR LAZY?						
HIGH AMOUNT OF DEAD BEES?						
QUEEN BEE IS PRESENT / IDENTIFIABLE?						
INFESTATION BY ANTS / ANTS PRESENT?						
INFESTATION BY WAX MOTH / WAX MOTH PRESENT?						
NEGATIVE ODOUR NOTICEABLE?						

COLONY NAME
DATE
TIME

WEATHER CONDITIONS						
	___					
	___	☐	☐	☐	☐	☐

INSPECTION						
HIVE NUMBER	1	2	3	4	5	6

PRODUCTIVITY & REPRODUCTION						
AMOUNT OF HONEY						
GENERAL POPULATION						
AMOUNT OF BROOD						
AMOUNT OF SPACE						

BEHAVIOUR & ACTIVITIES						
USUAL ENTERING AND EXITING ACTIVITY?						
CALM BEHAVIOUR WHEN OPENING HIVE?						
BEES BRINGING POLLEN INTO HIVE?						
SIGNS OF ROBBERY AMONG THE BEES?						

HEALTH STATUS						
BEES SEEM WEAK OR LAZY?						
HIGH AMOUNT OF DEAD BEES?						
QUEEN BEE IS PRESENT / IDENTIFIABLE?						
INFESTATION BY ANTS / ANTS PRESENT?						
INFESTATION BY WAX MOTH / WAX MOTH PRESENT?						
NEGATIVE ODOUR NOTICEABLE?						

COLONY NAME	
DATE	
TIME	

WEATHER CONDITIONS						
	___					
	___	☐	☐	☐	☐	☐

INSPECTION						
HIVE NUMBER	1	2	3	4	5	6

PRODUCTIVITY & REPRODUCTION						
AMOUNT OF HONEY						
GENERAL POPULATION						
AMOUNT OF BROOD						
AMOUNT OF SPACE						

BEHAVIOUR & ACTIVITIES						
USUAL ENTERING AND EXITING ACTIVITY?						
CALM BEHAVIOUR WHEN OPENING HIVE?						
BEES BRINGING POLLEN INTO HIVE?						
SIGNS OF ROBBERY AMONG THE BEES?						

HEALTH STATUS						
BEES SEEM WEAK OR LAZY?						
HIGH AMOUNT OF DEAD BEES?						
QUEEN BEE IS PRESENT / IDENTIFIABLE?						
INFESTATION BY ANTS / ANTS PRESENT?						
INFESTATION BY WAX MOTH / WAX MOTH PRESENT?						
NEGATIVE ODOUR NOTICEABLE?						

COLONY NAME	
DATE	
TIME	

WEATHER CONDITIONS						
Temperature	___	☐ Sunny	☐ Partly cloudy	☐ Rain	☐ Storm	☐ Snow
Wind	___					

INSPECTION

HIVE NUMBER	1	2	3	4	5	6

PRODUCTIVITY & REPRODUCTION

AMOUNT OF HONEY						
GENERAL POPULATION						
AMOUNT OF BROOD						
AMOUNT OF SPACE						

BEHAVIOUR & ACTIVITIES

USUAL ENTERING AND EXITING ACTIVITY?						
CALM BEHAVIOUR WHEN OPENING HIVE?						
BEES BRINGING POLLEN INTO HIVE?						
SIGNS OF ROBBERY AMONG THE BEES?						

HEALTH STATUS

BEES SEEM WEAK OR LAZY?						
HIGH AMOUNT OF DEAD BEES?						
QUEEN BEE IS PRESENT / IDENTIFIABLE?						
INFESTATION BY ANTS / ANTS PRESENT?						
INFESTATION BY WAX MOTH / WAX MOTH PRESENT?						
NEGATIVE ODOUR NOTICEABLE?						

COLONY NAME	
DATE	
TIME	

WEATHER CONDITIONS						
Temperature	___	☐	☐	☐	☐	☐
Wind	___					

INSPECTION						
HIVE NUMBER	1	2	3	4	5	6

PRODUCTIVITY & REPRODUCTION						
AMOUNT OF HONEY						
GENERAL POPULATION						
AMOUNT OF BROOD						
AMOUNT OF SPACE						

BEHAVIOUR & ACTIVITIES						
USUAL ENTERING AND EXITING ACTIVITY?						
CALM BEHAVIOUR WHEN OPENING HIVE?						
BEES BRINGING POLLEN INTO HIVE?						
SIGNS OF ROBBERY AMONG THE BEES?						

HEALTH STATUS						
BEES SEEM WEAK OR LAZY?						
HIGH AMOUNT OF DEAD BEES?						
QUEEN BEE IS PRESENT / IDENTIFIABLE?						
INFESTATION BY ANTS / ANTS PRESENT?						
INFESTATION BY WAX MOTH / WAX MOTH PRESENT?						
NEGATIVE ODOUR NOTICEABLE?						

COLONY NAME	
DATE	
TIME	

WEATHER CONDITIONS					
___	☐	☐	☐	☐	☐
___					

INSPECTION

HIVE NUMBER	1	2	3	4	5	6

PRODUCTIVITY & REPRODUCTION

AMOUNT OF HONEY						
GENERAL POPULATION						
AMOUNT OF BROOD						
AMOUNT OF SPACE						

BEHAVIOUR & ACTIVITIES

USUAL ENTERING AND EXITING ACTIVITY?						
CALM BEHAVIOUR WHEN OPENING HIVE?						
BEES BRINGING POLLEN INTO HIVE?						
SIGNS OF ROBBERY AMONG THE BEES?						

HEALTH STATUS

BEES SEEM WEAK OR LAZY?						
HIGH AMOUNT OF DEAD BEES?						
QUEEN BEE IS PRESENT / IDENTIFIABLE?						
INFESTATION BY ANTS / ANTS PRESENT?						
INFESTATION BY WAX MOTH / WAX MOTH PRESENT?						
NEGATIVE ODOUR NOTICEABLE?						

COLONY NAME	
DATE	
TIME	

WEATHER CONDITIONS						
Temperature	___	Sunny	Partly cloudy	Rain	Storm	Snow
Wind	___	☐	☐	☐	☐	☐

INSPECTION

HIVE NUMBER	1	2	3	4	5	6

PRODUCTIVITY & REPRODUCTION

AMOUNT OF HONEY						
GENERAL POPULATION						
AMOUNT OF BROOD						
AMOUNT OF SPACE						

BEHAVIOUR & ACTIVITIES

USUAL ENTERING AND EXITING ACTIVITY?						
CALM BEHAVIOUR WHEN OPENING HIVE?						
BEES BRINGING POLLEN INTO HIVE?						
SIGNS OF ROBBERY AMONG THE BEES?						

HEALTH STATUS

BEES SEEM WEAK OR LAZY?						
HIGH AMOUNT OF DEAD BEES?						
QUEEN BEE IS PRESENT / IDENTIFIABLE?						
INFESTATION BY ANTS / ANTS PRESENT?						
INFESTATION BY WAX MOTH / WAX MOTH PRESENT?						
NEGATIVE ODOUR NOTICEABLE?						

COLONY NAME
DATE
TIME

WEATHER CONDITIONS						
	___					
	___	☐	☐	☐	☐	☐

INSPECTION

HIVE NUMBER	1	2	3	4	5	6

PRODUCTIVITY & REPRODUCTION

AMOUNT OF HONEY						
GENERAL POPULATION						
AMOUNT OF BROOD						
AMOUNT OF SPACE						

BEHAVIOUR & ACTIVITIES

USUAL ENTERING AND EXITING ACTIVITY?						
CALM BEHAVIOUR WHEN OPENING HIVE?						
BEES BRINGING POLLEN INTO HIVE?						
SIGNS OF ROBBERY AMONG THE BEES?						

HEALTH STATUS

BEES SEEM WEAK OR LAZY?						
HIGH AMOUNT OF DEAD BEES?						
QUEEN BEE IS PRESENT / IDENTIFIABLE?						
INFESTATION BY ANTS / ANTS PRESENT?						
INFESTATION BY WAX MOTH / WAX MOTH PRESENT?						
NEGATIVE ODOUR NOTICEABLE?						

COLONY NAME	
DATE	
TIME	

WEATHER CONDITIONS						
Temperature	___	Sunny	Partly cloudy	Rain	Storm	Snow
Wind	___	☐	☐	☐	☐	☐

INSPECTION						
HIVE NUMBER	1	2	3	4	5	6

PRODUCTIVITY & REPRODUCTION						
AMOUNT OF HONEY						
GENERAL POPULATION						
AMOUNT OF BROOD						
AMOUNT OF SPACE						

BEHAVIOUR & ACTIVITIES						
USUAL ENTERING AND EXITING ACTIVITY?						
CALM BEHAVIOUR WHEN OPENING HIVE?						
BEES BRINGING POLLEN INTO HIVE?						
SIGNS OF ROBBERY AMONG THE BEES?						

HEALTH STATUS						
BEES SEEM WEAK OR LAZY?						
HIGH AMOUNT OF DEAD BEES?						
QUEEN BEE IS PRESENT / IDENTIFIABLE?						
INFESTATION BY ANTS / ANTS PRESENT?						
INFESTATION BY WAX MOTH / WAX MOTH PRESENT?						
NEGATIVE ODOUR NOTICEABLE?						

COLONY NAME
DATE
TIME

WEATHER CONDITIONS						
___						
___	☐	☐	☐	☐	☐	

INSPECTION

HIVE NUMBER	1	2	3	4	5	6

PRODUCTIVITY & REPRODUCTION

AMOUNT OF HONEY						
GENERAL POPULATION						
AMOUNT OF BROOD						
AMOUNT OF SPACE						

BEHAVIOUR & ACTIVITIES

USUAL ENTERING AND EXITING ACTIVITY?						
CALM BEHAVIOUR WHEN OPENING HIVE?						
BEES BRINGING POLLEN INTO HIVE?						
SIGNS OF ROBBERY AMONG THE BEES?						

HEALTH STATUS

BEES SEEM WEAK OR LAZY?						
HIGH AMOUNT OF DEAD BEES?						
QUEEN BEE IS PRESENT / IDENTIFIABLE?						
INFESTATION BY ANTS / ANTS PRESENT?						
INFESTATION BY WAX MOTH / WAX MOTH PRESENT?						
NEGATIVE ODOUR NOTICEABLE?						

COLONY NAME	
DATE	
TIME	

WEATHER CONDITIONS						
Temperature	___	☐	☐	☐	☐	☐
Wind	___					

INSPECTION						
HIVE NUMBER	1	2	3	4	5	6

PRODUCTIVITY & REPRODUCTION						
AMOUNT OF HONEY						
GENERAL POPULATION						
AMOUNT OF BROOD						
AMOUNT OF SPACE						

BEHAVIOUR & ACTIVITIES						
USUAL ENTERING AND EXITING ACTIVITY?						
CALM BEHAVIOUR WHEN OPENING HIVE?						
BEES BRINGING POLLEN INTO HIVE?						
SIGNS OF ROBBERY AMONG THE BEES?						

HEALTH STATUS						
BEES SEEM WEAK OR LAZY?						
HIGH AMOUNT OF DEAD BEES?						
QUEEN BEE IS PRESENT / IDENTIFIABLE?						
INFESTATION BY ANTS / ANTS PRESENT?						
INFESTATION BY WAX MOTH / WAX MOTH PRESENT?						
NEGATIVE ODOUR NOTICEABLE?						

COLONY NAME	
DATE	
TIME	

WEATHER CONDITIONS						
___						
___	☐	☐	☐	☐	☐	

INSPECTION						
HIVE NUMBER	1	2	3	4	5	6

PRODUCTIVITY & REPRODUCTION						
AMOUNT OF HONEY						
GENERAL POPULATION						
AMOUNT OF BROOD						
AMOUNT OF SPACE						

BEHAVIOUR & ACTIVITIES						
USUAL ENTERING AND EXITING ACTIVITY?						
CALM BEHAVIOUR WHEN OPENING HIVE?						
BEES BRINGING POLLEN INTO HIVE?						
SIGNS OF ROBBERY AMONG THE BEES?						

HEALTH STATUS						
BEES SEEM WEAK OR LAZY?						
HIGH AMOUNT OF DEAD BEES?						
QUEEN BEE IS PRESENT / IDENTIFIABLE?						
INFESTATION BY ANTS / ANTS PRESENT?						
INFESTATION BY WAX MOTH / WAX MOTH PRESENT?						
NEGATIVE ODOUR NOTICEABLE?						

COLONY NAME	
DATE	
TIME	

WEATHER CONDITIONS						
Temperature	___	☐	☐	☐	☐	☐
Wind	___					

INSPECTION						
HIVE NUMBER	1	2	3	4	5	6

PRODUCTIVITY & REPRODUCTION						
AMOUNT OF HONEY						
GENERAL POPULATION						
AMOUNT OF BROOD						
AMOUNT OF SPACE						

BEHAVIOUR & ACTIVITIES						
USUAL ENTERING AND EXITING ACTIVITY?						
CALM BEHAVIOUR WHEN OPENING HIVE?						
BEES BRINGING POLLEN INTO HIVE?						
SIGNS OF ROBBERY AMONG THE BEES?						

HEALTH STATUS						
BEES SEEM WEAK OR LAZY?						
HIGH AMOUNT OF DEAD BEES?						
QUEEN BEE IS PRESENT / IDENTIFIABLE?						
INFESTATION BY ANTS / ANTS PRESENT?						
INFESTATION BY WAX MOTH / WAX MOTH PRESENT?						
NEGATIVE ODOUR NOTICEABLE?						

COLONY NAME
DATE
TIME

WEATHER CONDITIONS						
	___					
	___	☐	☐	☐	☐	☐

INSPECTION						
HIVE NUMBER	1	2	3	4	5	6

PRODUCTIVITY & REPRODUCTION						
AMOUNT OF HONEY						
GENERAL POPULATION						
AMOUNT OF BROOD						
AMOUNT OF SPACE						

BEHAVIOUR & ACTIVITIES						
USUAL ENTERING AND EXITING ACTIVITY?						
CALM BEHAVIOUR WHEN OPENING HIVE?						
BEES BRINGING POLLEN INTO HIVE?						
SIGNS OF ROBBERY AMONG THE BEES?						

HEALTH STATUS						
BEES SEEM WEAK OR LAZY?						
HIGH AMOUNT OF DEAD BEES?						
QUEEN BEE IS PRESENT / IDENTIFIABLE?						
INFESTATION BY ANTS / ANTS PRESENT?						
INFESTATION BY WAX MOTH / WAX MOTH PRESENT?						
NEGATIVE ODOUR NOTICEABLE?						

COLONY NAME	
DATE	
TIME	

WEATHER CONDITIONS						
Temperature	___	Sunny	Partly cloudy	Rain	Storm	Snow
Wind	___	☐	☐	☐	☐	☐

INSPECTION

HIVE NUMBER	1	2	3	4	5	6

PRODUCTIVITY & REPRODUCTION

AMOUNT OF HONEY						
GENERAL POPULATION						
AMOUNT OF BROOD						
AMOUNT OF SPACE						

BEHAVIOUR & ACTIVITIES

USUAL ENTERING AND EXITING ACTIVITY?						
CALM BEHAVIOUR WHEN OPENING HIVE?						
BEES BRINGING POLLEN INTO HIVE?						
SIGNS OF ROBBERY AMONG THE BEES?						

HEALTH STATUS

BEES SEEM WEAK OR LAZY?						
HIGH AMOUNT OF DEAD BEES?						
QUEEN BEE IS PRESENT / IDENTIFIABLE?						
INFESTATION BY ANTS / ANTS PRESENT?						
INFESTATION BY WAX MOTH / WAX MOTH PRESENT?						
NEGATIVE ODOUR NOTICEABLE?						

COLONY NAME	
DATE	
TIME	

WEATHER CONDITIONS						
Temperature	___	Sunny	Partly cloudy	Rain	Thunderstorm	Snow
Wind	___	☐	☐	☐	☐	☐

INSPECTION						
HIVE NUMBER	1	2	3	4	5	6

PRODUCTIVITY & REPRODUCTION						
AMOUNT OF HONEY						
GENERAL POPULATION						
AMOUNT OF BROOD						
AMOUNT OF SPACE						

BEHAVIOUR & ACTIVITIES						
USUAL ENTERING AND EXITING ACTIVITY?						
CALM BEHAVIOUR WHEN OPENING HIVE?						
BEES BRINGING POLLEN INTO HIVE?						
SIGNS OF ROBBERY AMONG THE BEES?						

HEALTH STATUS						
BEES SEEM WEAK OR LAZY?						
HIGH AMOUNT OF DEAD BEES?						
QUEEN BEE IS PRESENT / IDENTIFIABLE?						
INFESTATION BY ANTS / ANTS PRESENT?						
INFESTATION BY WAX MOTH / WAX MOTH PRESENT?						
NEGATIVE ODOUR NOTICEABLE?						

COLONY NAME	
DATE	
TIME	

WEATHER CONDITIONS						
	___					
	___	☐	☐	☐	☐	☐

INSPECTION						
HIVE NUMBER	1	2	3	4	5	6

PRODUCTIVITY & REPRODUCTION						
AMOUNT OF HONEY						
GENERAL POPULATION						
AMOUNT OF BROOD						
AMOUNT OF SPACE						

BEHAVIOUR & ACTIVITIES						
USUAL ENTERING AND EXITING ACTIVITY?						
CALM BEHAVIOUR WHEN OPENING HIVE?						
BEES BRINGING POLLEN INTO HIVE?						
SIGNS OF ROBBERY AMONG THE BEES?						

HEALTH STATUS						
BEES SEEM WEAK OR LAZY?						
HIGH AMOUNT OF DEAD BEES?						
QUEEN BEE IS PRESENT / IDENTIFIABLE?						
INFESTATION BY ANTS / ANTS PRESENT?						
INFESTATION BY WAX MOTH / WAX MOTH PRESENT?						
NEGATIVE ODOUR NOTICEABLE?						

COLONY NAME	
DATE	
TIME	

WEATHER CONDITIONS						
Temperature	___	☐	☐	☐	☐	☐
Wind	___					

INSPECTION						
HIVE NUMBER	1	2	3	4	5	6

PRODUCTIVITY & REPRODUCTION						
AMOUNT OF HONEY						
GENERAL POPULATION						
AMOUNT OF BROOD						
AMOUNT OF SPACE						

BEHAVIOUR & ACTIVITIES						
USUAL ENTERING AND EXITING ACTIVITY?						
CALM BEHAVIOUR WHEN OPENING HIVE?						
BEES BRINGING POLLEN INTO HIVE?						
SIGNS OF ROBBERY AMONG THE BEES?						

HEALTH STATUS						
BEES SEEM WEAK OR LAZY?						
HIGH AMOUNT OF DEAD BEES?						
QUEEN BEE IS PRESENT / IDENTIFIABLE?						
INFESTATION BY ANTS / ANTS PRESENT?						
INFESTATION BY WAX MOTH / WAX MOTH PRESENT?						
NEGATIVE ODOUR NOTICEABLE?						

COLONY NAME	
DATE	
TIME	

WEATHER CONDITIONS					
___					
___	☐	☐	☐	☐	☐

INSPECTION						
HIVE NUMBER	1	2	3	4	5	6

PRODUCTIVITY & REPRODUCTION						
AMOUNT OF HONEY						
GENERAL POPULATION						
AMOUNT OF BROOD						
AMOUNT OF SPACE						

BEHAVIOUR & ACTIVITIES						
USUAL ENTERING AND EXITING ACTIVITY?						
CALM BEHAVIOUR WHEN OPENING HIVE?						
BEES BRINGING POLLEN INTO HIVE?						
SIGNS OF ROBBERY AMONG THE BEES?						

HEALTH STATUS						
BEES SEEM WEAK OR LAZY?						
HIGH AMOUNT OF DEAD BEES?						
QUEEN BEE IS PRESENT / IDENTIFIABLE?						
INFESTATION BY ANTS / ANTS PRESENT?						
INFESTATION BY WAX MOTH / WAX MOTH PRESENT?						
NEGATIVE ODOUR NOTICEABLE?						

COLONY NAME	
DATE	
TIME	

WEATHER CONDITIONS						
Temperature	___	Sunny	Partly cloudy	Rain	Storm	Snow
Wind	___	☐	☐	☐	☐	☐

INSPECTION						
HIVE NUMBER	1	2	3	4	5	6

PRODUCTIVITY & REPRODUCTION						
AMOUNT OF HONEY						
GENERAL POPULATION						
AMOUNT OF BROOD						
AMOUNT OF SPACE						

BEHAVIOUR & ACTIVITIES						
USUAL ENTERING AND EXITING ACTIVITY?						
CALM BEHAVIOUR WHEN OPENING HIVE?						
BEES BRINGING POLLEN INTO HIVE?						
SIGNS OF ROBBERY AMONG THE BEES?						

HEALTH STATUS						
BEES SEEM WEAK OR LAZY?						
HIGH AMOUNT OF DEAD BEES?						
QUEEN BEE IS PRESENT / IDENTIFIABLE?						
INFESTATION BY ANTS / ANTS PRESENT?						
INFESTATION BY WAX MOTH / WAX MOTH PRESENT?						
NEGATIVE ODOUR NOTICEABLE?						

COLONY NAME	
DATE	
TIME	

WEATHER CONDITIONS						
Temperature	___	☐	☐	☐	☐	☐
Wind	___	☐	☐	☐	☐	☐

INSPECTION						
HIVE NUMBER	1	2	3	4	5	6

PRODUCTIVITY & REPRODUCTION						
AMOUNT OF HONEY						
GENERAL POPULATION						
AMOUNT OF BROOD						
AMOUNT OF SPACE						

BEHAVIOUR & ACTIVITIES						
USUAL ENTERING AND EXITING ACTIVITY?						
CALM BEHAVIOUR WHEN OPENING HIVE?						
BEES BRINGING POLLEN INTO HIVE?						
SIGNS OF ROBBERY AMONG THE BEES?						

HEALTH STATUS						
BEES SEEM WEAK OR LAZY?						
HIGH AMOUNT OF DEAD BEES?						
QUEEN BEE IS PRESENT / IDENTIFIABLE?						
INFESTATION BY ANTS / ANTS PRESENT?						
INFESTATION BY WAX MOTH / WAX MOTH PRESENT?						
NEGATIVE ODOUR NOTICEABLE?						

COLONY NAME	
DATE	
TIME	

WEATHER CONDITIONS						
Temperature	___	☐	☐	☐	☐	☐
Wind	___	☐	☐	☐	☐	☐

INSPECTION

HIVE NUMBER	1	2	3	4	5	6

PRODUCTIVITY & REPRODUCTION

AMOUNT OF HONEY						
GENERAL POPULATION						
AMOUNT OF BROOD						
AMOUNT OF SPACE						

BEHAVIOUR & ACTIVITIES

USUAL ENTERING AND EXITING ACTIVITY?						
CALM BEHAVIOUR WHEN OPENING HIVE?						
BEES BRINGING POLLEN INTO HIVE?						
SIGNS OF ROBBERY AMONG THE BEES?						

HEALTH STATUS

BEES SEEM WEAK OR LAZY?						
HIGH AMOUNT OF DEAD BEES?						
QUEEN BEE IS PRESENT / IDENTIFIABLE?						
INFESTATION BY ANTS / ANTS PRESENT?						
INFESTATION BY WAX MOTH / WAX MOTH PRESENT?						
NEGATIVE ODOUR NOTICEABLE?						

COLONY NAME	
DATE	
TIME	

WEATHER CONDITIONS					
___					
___	☐	☐	☐	☐	☐

INSPECTION						
HIVE NUMBER	1	2	3	4	5	6

PRODUCTIVITY & REPRODUCTION						
AMOUNT OF HONEY						
GENERAL POPULATION						
AMOUNT OF BROOD						
AMOUNT OF SPACE						

BEHAVIOUR & ACTIVITIES						
USUAL ENTERING AND EXITING ACTIVITY?						
CALM BEHAVIOUR WHEN OPENING HIVE?						
BEES BRINGING POLLEN INTO HIVE?						
SIGNS OF ROBBERY AMONG THE BEES?						

HEALTH STATUS						
BEES SEEM WEAK OR LAZY?						
HIGH AMOUNT OF DEAD BEES?						
QUEEN BEE IS PRESENT / IDENTIFIABLE?						
INFESTATION BY ANTS / ANTS PRESENT?						
INFESTATION BY WAX MOTH / WAX MOTH PRESENT?						
NEGATIVE ODOUR NOTICEABLE?						

COLONY NAME	
DATE	
TIME	

WEATHER CONDITIONS						
Temperature	___	☐	☐	☐	☐	☐
Wind	___	☐	☐	☐	☐	☐

INSPECTION						
HIVE NUMBER	1	2	3	4	5	6

PRODUCTIVITY & REPRODUCTION						
AMOUNT OF HONEY						
GENERAL POPULATION						
AMOUNT OF BROOD						
AMOUNT OF SPACE						

BEHAVIOUR & ACTIVITIES						
USUAL ENTERING AND EXITING ACTIVITY?						
CALM BEHAVIOUR WHEN OPENING HIVE?						
BEES BRINGING POLLEN INTO HIVE?						
SIGNS OF ROBBERY AMONG THE BEES?						

HEALTH STATUS						
BEES SEEM WEAK OR LAZY?						
HIGH AMOUNT OF DEAD BEES?						
QUEEN BEE IS PRESENT / IDENTIFIABLE?						
INFESTATION BY ANTS / ANTS PRESENT?						
INFESTATION BY WAX MOTH / WAX MOTH PRESENT?						
NEGATIVE ODOUR NOTICEABLE?						

COLONY NAME	
DATE	
TIME	

WEATHER CONDITIONS						
Temperature	___	Sunny	Partly cloudy	Rain	Storm	Snow
Wind	___	☐	☐	☐	☐	☐

INSPECTION						
HIVE NUMBER	1	2	3	4	5	6

PRODUCTIVITY & REPRODUCTION						
AMOUNT OF HONEY						
GENERAL POPULATION						
AMOUNT OF BROOD						
AMOUNT OF SPACE						

BEHAVIOUR & ACTIVITIES						
USUAL ENTERING AND EXITING ACTIVITY?						
CALM BEHAVIOUR WHEN OPENING HIVE?						
BEES BRINGING POLLEN INTO HIVE?						
SIGNS OF ROBBERY AMONG THE BEES?						

HEALTH STATUS						
BEES SEEM WEAK OR LAZY?						
HIGH AMOUNT OF DEAD BEES?						
QUEEN BEE IS PRESENT / IDENTIFIABLE?						
INFESTATION BY ANTS / ANTS PRESENT?						
INFESTATION BY WAX MOTH / WAX MOTH PRESENT?						
NEGATIVE ODOUR NOTICEABLE?						

COLONY NAME	
DATE	
TIME	

WEATHER CONDITIONS						
	____					
	____	☐	☐	☐	☐	☐

INSPECTION						
HIVE NUMBER	1	2	3	4	5	6

PRODUCTIVITY & REPRODUCTION						
AMOUNT OF HONEY						
GENERAL POPULATION						
AMOUNT OF BROOD						
AMOUNT OF SPACE						

BEHAVIOUR & ACTIVITIES						
USUAL ENTERING AND EXITING ACTIVITY?						
CALM BEHAVIOUR WHEN OPENING HIVE?						
BEES BRINGING POLLEN INTO HIVE?						
SIGNS OF ROBBERY AMONG THE BEES?						

HEALTH STATUS						
BEES SEEM WEAK OR LAZY?						
HIGH AMOUNT OF DEAD BEES?						
QUEEN BEE IS PRESENT / IDENTIFIABLE?						
INFESTATION BY ANTS / ANTS PRESENT?						
INFESTATION BY WAX MOTH / WAX MOTH PRESENT?						
NEGATIVE ODOUR NOTICEABLE?						

COLONY NAME	
DATE	
TIME	

WEATHER CONDITIONS						
(temperature)	___	(sunny)	(partly cloudy)	(rain)	(storm)	(snow)
(wind)	___	☐	☐	☐	☐	☐

INSPECTION

HIVE NUMBER	1	2	3	4	5	6

PRODUCTIVITY & REPRODUCTION

	1	2	3	4	5	6
AMOUNT OF HONEY						
GENERAL POPULATION						
AMOUNT OF BROOD						
AMOUNT OF SPACE						

BEHAVIOUR & ACTIVITIES

	1	2	3	4	5	6
USUAL ENTERING AND EXITING ACTIVITY?						
CALM BEHAVIOUR WHEN OPENING HIVE?						
BEES BRINGING POLLEN INTO HIVE?						
SIGNS OF ROBBERY AMONG THE BEES?						

HEALTH STATUS

	1	2	3	4	5	6
BEES SEEM WEAK OR LAZY?						
HIGH AMOUNT OF DEAD BEES?						
QUEEN BEE IS PRESENT / IDENTIFIABLE?						
INFESTATION BY ANTS / ANTS PRESENT?						
INFESTATION BY WAX MOTH / WAX MOTH PRESENT?						
NEGATIVE ODOUR NOTICEABLE?						

COLONY NAME	
DATE	
TIME	

WEATHER CONDITIONS						
Temperature	___					
Wind	___	☐	☐	☐	☐	☐

INSPECTION						
HIVE NUMBER	1	2	3	4	5	6

PRODUCTIVITY & REPRODUCTION						
AMOUNT OF HONEY						
GENERAL POPULATION						
AMOUNT OF BROOD						
AMOUNT OF SPACE						

BEHAVIOUR & ACTIVITIES						
USUAL ENTERING AND EXITING ACTIVITY?						
CALM BEHAVIOUR WHEN OPENING HIVE?						
BEES BRINGING POLLEN INTO HIVE?						
SIGNS OF ROBBERY AMONG THE BEES?						

HEALTH STATUS						
BEES SEEM WEAK OR LAZY?						
HIGH AMOUNT OF DEAD BEES?						
QUEEN BEE IS PRESENT / IDENTIFIABLE?						
INFESTATION BY ANTS / ANTS PRESENT?						
INFESTATION BY WAX MOTH / WAX MOTH PRESENT?						
NEGATIVE ODOUR NOTICEABLE?						

COLONY NAME	
DATE	
TIME	

WEATHER CONDITIONS						
(temperature)	___					
(wind)	___	☐	☐	☐	☐	☐

INSPECTION						
HIVE NUMBER	1	2	3	4	5	6

PRODUCTIVITY & REPRODUCTION						
AMOUNT OF HONEY						
GENERAL POPULATION						
AMOUNT OF BROOD						
AMOUNT OF SPACE						

BEHAVIOUR & ACTIVITIES						
USUAL ENTERING AND EXITING ACTIVITY?						
CALM BEHAVIOUR WHEN OPENING HIVE?						
BEES BRINGING POLLEN INTO HIVE?						
SIGNS OF ROBBERY AMONG THE BEES?						

HEALTH STATUS						
BEES SEEM WEAK OR LAZY?						
HIGH AMOUNT OF DEAD BEES?						
QUEEN BEE IS PRESENT / IDENTIFIABLE?						
INFESTATION BY ANTS / ANTS PRESENT?						
INFESTATION BY WAX MOTH / WAX MOTH PRESENT?						
NEGATIVE ODOUR NOTICEABLE?						

COLONY NAME	
DATE	
TIME	

WEATHER CONDITIONS						
Temperature	___	☐	☐	☐	☐	☐
Wind	___	☐	☐	☐	☐	☐

INSPECTION

HIVE NUMBER	1	2	3	4	5	6

PRODUCTIVITY & REPRODUCTION

AMOUNT OF HONEY						
GENERAL POPULATION						
AMOUNT OF BROOD						
AMOUNT OF SPACE						

BEHAVIOUR & ACTIVITIES

USUAL ENTERING AND EXITING ACTIVITY?						
CALM BEHAVIOUR WHEN OPENING HIVE?						
BEES BRINGING POLLEN INTO HIVE?						
SIGNS OF ROBBERY AMONG THE BEES?						

HEALTH STATUS

BEES SEEM WEAK OR LAZY?						
HIGH AMOUNT OF DEAD BEES?						
QUEEN BEE IS PRESENT / IDENTIFIABLE?						
INFESTATION BY ANTS / ANTS PRESENT?						
INFESTATION BY WAX MOTH / WAX MOTH PRESENT?						
NEGATIVE ODOUR NOTICEABLE?						

COLONY NAME	
DATE	
TIME	

WEATHER CONDITIONS						
Temperature	___	☐	☐	☐	☐	☐
Wind	___					

INSPECTION

HIVE NUMBER	1	2	3	4	5	6

PRODUCTIVITY & REPRODUCTION

	1	2	3	4	5	6
AMOUNT OF HONEY						
GENERAL POPULATION						
AMOUNT OF BROOD						
AMOUNT OF SPACE						

BEHAVIOUR & ACTIVITIES

	1	2	3	4	5	6
USUAL ENTERING AND EXITING ACTIVITY?						
CALM BEHAVIOUR WHEN OPENING HIVE?						
BEES BRINGING POLLEN INTO HIVE?						
SIGNS OF ROBBERY AMONG THE BEES?						

HEALTH STATUS

	1	2	3	4	5	6
BEES SEEM WEAK OR LAZY?						
HIGH AMOUNT OF DEAD BEES?						
QUEEN BEE IS PRESENT / IDENTIFIABLE?						
INFESTATION BY ANTS / ANTS PRESENT?						
INFESTATION BY WAX MOTH / WAX MOTH PRESENT?						
NEGATIVE ODOUR NOTICEABLE?						

COLONY NAME	
DATE	
TIME	

WEATHER CONDITIONS						
Temperature	___	Sunny	Partly cloudy	Rain	Storm	Snow
Wind	___	☐	☐	☐	☐	☐

INSPECTION

HIVE NUMBER	1	2	3	4	5	6

PRODUCTIVITY & REPRODUCTION

AMOUNT OF HONEY						
GENERAL POPULATION						
AMOUNT OF BROOD						
AMOUNT OF SPACE						

BEHAVIOUR & ACTIVITIES

USUAL ENTERING AND EXITING ACTIVITY?						
CALM BEHAVIOUR WHEN OPENING HIVE?						
BEES BRINGING POLLEN INTO HIVE?						
SIGNS OF ROBBERY AMONG THE BEES?						

HEALTH STATUS

BEES SEEM WEAK OR LAZY?						
HIGH AMOUNT OF DEAD BEES?						
QUEEN BEE IS PRESENT / IDENTIFIABLE?						
INFESTATION BY ANTS / ANTS PRESENT?						
INFESTATION BY WAX MOTH / WAX MOTH PRESENT?						
NEGATIVE ODOUR NOTICEABLE?						

COLONY NAME	
DATE	
TIME	

WEATHER CONDITIONS						
Temperature	___	☐	☐	☐	☐	☐
Wind	___	☐	☐	☐	☐	☐

INSPECTION

HIVE NUMBER	1	2	3	4	5	6

PRODUCTIVITY & REPRODUCTION

AMOUNT OF HONEY						
GENERAL POPULATION						
AMOUNT OF BROOD						
AMOUNT OF SPACE						

BEHAVIOUR & ACTIVITIES

USUAL ENTERING AND EXITING ACTIVITY?						
CALM BEHAVIOUR WHEN OPENING HIVE?						
BEES BRINGING POLLEN INTO HIVE?						
SIGNS OF ROBBERY AMONG THE BEES?						

HEALTH STATUS

BEES SEEM WEAK OR LAZY?						
HIGH AMOUNT OF DEAD BEES?						
QUEEN BEE IS PRESENT / IDENTIFIABLE?						
INFESTATION BY ANTS / ANTS PRESENT?						
INFESTATION BY WAX MOTH / WAX MOTH PRESENT?						
NEGATIVE ODOUR NOTICEABLE?						

COLONY NAME	
DATE	
TIME	

WEATHER CONDITIONS						
___						
___	☐	☐	☐	☐	☐	

INSPECTION						
HIVE NUMBER	1	2	3	4	5	6

PRODUCTIVITY & REPRODUCTION						
AMOUNT OF HONEY						
GENERAL POPULATION						
AMOUNT OF BROOD						
AMOUNT OF SPACE						

BEHAVIOUR & ACTIVITIES						
USUAL ENTERING AND EXITING ACTIVITY?						
CALM BEHAVIOUR WHEN OPENING HIVE?						
BEES BRINGING POLLEN INTO HIVE?						
SIGNS OF ROBBERY AMONG THE BEES?						

HEALTH STATUS						
BEES SEEM WEAK OR LAZY?						
HIGH AMOUNT OF DEAD BEES?						
QUEEN BEE IS PRESENT / IDENTIFIABLE?						
INFESTATION BY ANTS / ANTS PRESENT?						
INFESTATION BY WAX MOTH / WAX MOTH PRESENT?						
NEGATIVE ODOUR NOTICEABLE?						

COLONY NAME	
DATE	
TIME	

WEATHER CONDITIONS						
Temperature	___	☐	☐	☐	☐	☐
Wind	___	☐	☐	☐	☐	☐

INSPECTION

HIVE NUMBER	1	2	3	4	5	6
PRODUCTIVITY & REPRODUCTION						
AMOUNT OF HONEY						
GENERAL POPULATION						
AMOUNT OF BROOD						
AMOUNT OF SPACE						
BEHAVIOUR & ACTIVITIES						
USUAL ENTERING AND EXITING ACTIVITY?						
CALM BEHAVIOUR WHEN OPENING HIVE?						
BEES BRINGING POLLEN INTO HIVE?						
SIGNS OF ROBBERY AMONG THE BEES?						
HEALTH STATUS						
BEES SEEM WEAK OR LAZY?						
HIGH AMOUNT OF DEAD BEES?						
QUEEN BEE IS PRESENT / IDENTIFIABLE?						
INFESTATION BY ANTS / ANTS PRESENT?						
INFESTATION BY WAX MOTH / WAX MOTH PRESENT?						
NEGATIVE ODOUR NOTICEABLE?						

COLONY NAME	
DATE	
TIME	

WEATHER CONDITIONS
___
___

INSPECTION						
HIVE NUMBER	1	2	3	4	5	6

PRODUCTIVITY & REPRODUCTION						
AMOUNT OF HONEY						
GENERAL POPULATION						
AMOUNT OF BROOD						
AMOUNT OF SPACE						

BEHAVIOUR & ACTIVITIES						
USUAL ENTERING AND EXITING ACTIVITY?						
CALM BEHAVIOUR WHEN OPENING HIVE?						
BEES BRINGING POLLEN INTO HIVE?						
SIGNS OF ROBBERY AMONG THE BEES?						

HEALTH STATUS						
BEES SEEM WEAK OR LAZY?						
HIGH AMOUNT OF DEAD BEES?						
QUEEN BEE IS PRESENT / IDENTIFIABLE?						
INFESTATION BY ANTS / ANTS PRESENT?						
INFESTATION BY WAX MOTH / WAX MOTH PRESENT?						
NEGATIVE ODOUR NOTICEABLE?						

COLONY NAME	
DATE	
TIME	

WEATHER CONDITIONS						
Temperature	___	Sunny	Partly cloudy	Rain	Thunderstorm	Snow
Wind	___	☐	☐	☐	☐	☐

INSPECTION						
HIVE NUMBER	1	2	3	4	5	6

PRODUCTIVITY & REPRODUCTION						
AMOUNT OF HONEY						
GENERAL POPULATION						
AMOUNT OF BROOD						
AMOUNT OF SPACE						

BEHAVIOUR & ACTIVITIES						
USUAL ENTERING AND EXITING ACTIVITY?						
CALM BEHAVIOUR WHEN OPENING HIVE?						
BEES BRINGING POLLEN INTO HIVE?						
SIGNS OF ROBBERY AMONG THE BEES?						

HEALTH STATUS						
BEES SEEM WEAK OR LAZY?						
HIGH AMOUNT OF DEAD BEES?						
QUEEN BEE IS PRESENT / IDENTIFIABLE?						
INFESTATION BY ANTS / ANTS PRESENT?						
INFESTATION BY WAX MOTH / WAX MOTH PRESENT?						
NEGATIVE ODOUR NOTICEABLE?						

COLONY NAME
DATE
TIME

WEATHER CONDITIONS						
	___					
	___	☐	☐	☐	☐	☐

INSPECTION						
HIVE NUMBER	1	2	3	4	5	6

PRODUCTIVITY & REPRODUCTION						
AMOUNT OF HONEY						
GENERAL POPULATION						
AMOUNT OF BROOD						
AMOUNT OF SPACE						

BEHAVIOUR & ACTIVITIES						
USUAL ENTERING AND EXITING ACTIVITY?						
CALM BEHAVIOUR WHEN OPENING HIVE?						
BEES BRINGING POLLEN INTO HIVE?						
SIGNS OF ROBBERY AMONG THE BEES?						

HEALTH STATUS						
BEES SEEM WEAK OR LAZY?						
HIGH AMOUNT OF DEAD BEES?						
QUEEN BEE IS PRESENT / IDENTIFIABLE?						
INFESTATION BY ANTS / ANTS PRESENT?						
INFESTATION BY WAX MOTH / WAX MOTH PRESENT?						
NEGATIVE ODOUR NOTICEABLE?						

COLONY NAME	
DATE	
TIME	

WEATHER CONDITIONS						
	___					
	___	☐	☐	☐	☐	☐

INSPECTION						
HIVE NUMBER	1	2	3	4	5	6

PRODUCTIVITY & REPRODUCTION						
AMOUNT OF HONEY						
GENERAL POPULATION						
AMOUNT OF BROOD						
AMOUNT OF SPACE						

BEHAVIOUR & ACTIVITIES						
USUAL ENTERING AND EXITING ACTIVITY?						
CALM BEHAVIOUR WHEN OPENING HIVE?						
BEES BRINGING POLLEN INTO HIVE?						
SIGNS OF ROBBERY AMONG THE BEES?						

HEALTH STATUS						
BEES SEEM WEAK OR LAZY?						
HIGH AMOUNT OF DEAD BEES?						
QUEEN BEE IS PRESENT / IDENTIFIABLE?						
INFESTATION BY ANTS / ANTS PRESENT?						
INFESTATION BY WAX MOTH / WAX MOTH PRESENT?						
NEGATIVE ODOUR NOTICEABLE?						

COLONY NAME	
DATE	
TIME	

WEATHER CONDITIONS					
___	☐	☐	☐	☐	☐
___					

INSPECTION						
HIVE NUMBER	1	2	3	4	5	6

PRODUCTIVITY & REPRODUCTION						
AMOUNT OF HONEY						
GENERAL POPULATION						
AMOUNT OF BROOD						
AMOUNT OF SPACE						

BEHAVIOUR & ACTIVITIES						
USUAL ENTERING AND EXITING ACTIVITY?						
CALM BEHAVIOUR WHEN OPENING HIVE?						
BEES BRINGING POLLEN INTO HIVE?						
SIGNS OF ROBBERY AMONG THE BEES?						

HEALTH STATUS						
BEES SEEM WEAK OR LAZY?						
HIGH AMOUNT OF DEAD BEES?						
QUEEN BEE IS PRESENT / IDENTIFIABLE?						
INFESTATION BY ANTS / ANTS PRESENT?						
INFESTATION BY WAX MOTH / WAX MOTH PRESENT?						
NEGATIVE ODOUR NOTICEABLE?						

COLONY NAME	
DATE	
TIME	

WEATHER CONDITIONS						
Temperature	___	☐	☐	☐	☐	☐
Wind	___					

INSPECTION						
HIVE NUMBER	1	2	3	4	5	6

PRODUCTIVITY & REPRODUCTION						
AMOUNT OF HONEY						
GENERAL POPULATION						
AMOUNT OF BROOD						
AMOUNT OF SPACE						

BEHAVIOUR & ACTIVITIES						
USUAL ENTERING AND EXITING ACTIVITY?						
CALM BEHAVIOUR WHEN OPENING HIVE?						
BEES BRINGING POLLEN INTO HIVE?						
SIGNS OF ROBBERY AMONG THE BEES?						

HEALTH STATUS						
BEES SEEM WEAK OR LAZY?						
HIGH AMOUNT OF DEAD BEES?						
QUEEN BEE IS PRESENT / IDENTIFIABLE?						
INFESTATION BY ANTS / ANTS PRESENT?						
INFESTATION BY WAX MOTH / WAX MOTH PRESENT?						
NEGATIVE ODOUR NOTICEABLE?						

COLONY NAME	
DATE	
TIME	

WEATHER CONDITIONS						
(temperature)	___	☐	☐	☐	☐	☐
(wind)	___					

INSPECTION						
HIVE NUMBER	1	2	3	4	5	6

PRODUCTIVITY & REPRODUCTION						
AMOUNT OF HONEY						
GENERAL POPULATION						
AMOUNT OF BROOD						
AMOUNT OF SPACE						

BEHAVIOUR & ACTIVITIES						
USUAL ENTERING AND EXITING ACTIVITY?						
CALM BEHAVIOUR WHEN OPENING HIVE?						
BEES BRINGING POLLEN INTO HIVE?						
SIGNS OF ROBBERY AMONG THE BEES?						

HEALTH STATUS						
BEES SEEM WEAK OR LAZY?						
HIGH AMOUNT OF DEAD BEES?						
QUEEN BEE IS PRESENT / IDENTIFIABLE?						
INFESTATION BY ANTS / ANTS PRESENT?						
INFESTATION BY WAX MOTH / WAX MOTH PRESENT?						
NEGATIVE ODOUR NOTICEABLE?						

COLONY NAME	
DATE	
TIME	

WEATHER CONDITIONS						
Temperature	___	☐	☐	☐	☐	☐
Wind	___					

INSPECTION						
HIVE NUMBER	1	2	3	4	5	6

PRODUCTIVITY & REPRODUCTION						
AMOUNT OF HONEY						
GENERAL POPULATION						
AMOUNT OF BROOD						
AMOUNT OF SPACE						

BEHAVIOUR & ACTIVITIES						
USUAL ENTERING AND EXITING ACTIVITY?						
CALM BEHAVIOUR WHEN OPENING HIVE?						
BEES BRINGING POLLEN INTO HIVE?						
SIGNS OF ROBBERY AMONG THE BEES?						

HEALTH STATUS						
BEES SEEM WEAK OR LAZY?						
HIGH AMOUNT OF DEAD BEES?						
QUEEN BEE IS PRESENT / IDENTIFIABLE?						
INFESTATION BY ANTS / ANTS PRESENT?						
INFESTATION BY WAX MOTH / WAX MOTH PRESENT?						
NEGATIVE ODOUR NOTICEABLE?						

COLONY NAME	
DATE	
TIME	

WEATHER CONDITIONS						
Temperature	___	☐	☐	☐	☐	☐
Wind	___					

INSPECTION						
HIVE NUMBER	1	2	3	4	5	6

PRODUCTIVITY & REPRODUCTION						
AMOUNT OF HONEY						
GENERAL POPULATION						
AMOUNT OF BROOD						
AMOUNT OF SPACE						

BEHAVIOUR & ACTIVITIES						
USUAL ENTERING AND EXITING ACTIVITY?						
CALM BEHAVIOUR WHEN OPENING HIVE?						
BEES BRINGING POLLEN INTO HIVE?						
SIGNS OF ROBBERY AMONG THE BEES?						

HEALTH STATUS						
BEES SEEM WEAK OR LAZY?						
HIGH AMOUNT OF DEAD BEES?						
QUEEN BEE IS PRESENT / IDENTIFIABLE?						
INFESTATION BY ANTS / ANTS PRESENT?						
INFESTATION BY WAX MOTH / WAX MOTH PRESENT?						
NEGATIVE ODOUR NOTICEABLE?						

COLONY NAME	
DATE	
TIME	

WEATHER CONDITIONS						
	___					
	___	☐	☐	☐	☐	☐

INSPECTION						
HIVE NUMBER	1	2	3	4	5	6

PRODUCTIVITY & REPRODUCTION						
AMOUNT OF HONEY						
GENERAL POPULATION						
AMOUNT OF BROOD						
AMOUNT OF SPACE						

BEHAVIOUR & ACTIVITIES						
USUAL ENTERING AND EXITING ACTIVITY?						
CALM BEHAVIOUR WHEN OPENING HIVE?						
BEES BRINGING POLLEN INTO HIVE?						
SIGNS OF ROBBERY AMONG THE BEES?						

HEALTH STATUS						
BEES SEEM WEAK OR LAZY?						
HIGH AMOUNT OF DEAD BEES?						
QUEEN BEE IS PRESENT / IDENTIFIABLE?						
INFESTATION BY ANTS / ANTS PRESENT?						
INFESTATION BY WAX MOTH / WAX MOTH PRESENT?						
NEGATIVE ODOUR NOTICEABLE?						

COLONY NAME
DATE
TIME

WEATHER CONDITIONS						
	___					
	___	☐	☐	☐	☐	☐

INSPECTION						
HIVE NUMBER	1	2	3	4	5	6

PRODUCTIVITY & REPRODUCTION						
AMOUNT OF HONEY						
GENERAL POPULATION						
AMOUNT OF BROOD						
AMOUNT OF SPACE						

BEHAVIOUR & ACTIVITIES						
USUAL ENTERING AND EXITING ACTIVITY?						
CALM BEHAVIOUR WHEN OPENING HIVE?						
BEES BRINGING POLLEN INTO HIVE?						
SIGNS OF ROBBERY AMONG THE BEES?						

HEALTH STATUS						
BEES SEEM WEAK OR LAZY?						
HIGH AMOUNT OF DEAD BEES?						
QUEEN BEE IS PRESENT / IDENTIFIABLE?						
INFESTATION BY ANTS / ANTS PRESENT?						
INFESTATION BY WAX MOTH / WAX MOTH PRESENT?						
NEGATIVE ODOUR NOTICEABLE?						

COLONY NAME	
DATE	
TIME	

WEATHER CONDITIONS						
(temperature)	___	☐	☐	☐	☐	☐
(wind)	___					

INSPECTION						
HIVE NUMBER	1	2	3	4	5	6

PRODUCTIVITY & REPRODUCTION						
AMOUNT OF HONEY						
GENERAL POPULATION						
AMOUNT OF BROOD						
AMOUNT OF SPACE						

BEHAVIOUR & ACTIVITIES						
USUAL ENTERING AND EXITING ACTIVITY?						
CALM BEHAVIOUR WHEN OPENING HIVE?						
BEES BRINGING POLLEN INTO HIVE?						
SIGNS OF ROBBERY AMONG THE BEES?						

HEALTH STATUS						
BEES SEEM WEAK OR LAZY?						
HIGH AMOUNT OF DEAD BEES?						
QUEEN BEE IS PRESENT / IDENTIFIABLE?						
INFESTATION BY ANTS / ANTS PRESENT?						
INFESTATION BY WAX MOTH / WAX MOTH PRESENT?						
NEGATIVE ODOUR NOTICEABLE?						

Notes

Notes

Printed in Great Britain
by Amazon